INTRODUCTION TO

Optical Engineering

SOLUTIONS MANUAL

FRANCIS T. S. YU

Pennsylvania State University

XIANGYANG YANG

University of New Orleans

CAMBRIDGE
UNIVERSITY PRESS

PUBLISHED BY THE PRESS SYNDICATE OF THE UNIVERSITY OF CAMBRIDGE
The Pitt Building, Trumpington Street, Cambridge CB2 1RP, United Kingdom

CAMBRIDGE UNIVERSITY PRESS
The Edinburgh Building, Cambridge CB2 2RU, United Kingdom
40 West 20th Street, New York, NY 10011-4211, USA
10 Stamford Road, Oakleigh, Melbourne 3166, Australia

First published 1997

Printed in the United States of America

ISBN 0-521-59580-0 paperback

TABLE OF CONTENTS

Chapter 1

1.1 a). According to the Snell's law of refraction [see Eq(1.9)]

$$\frac{\sin\theta_2}{\sin\theta_1} = \frac{\eta_1}{\eta_2}$$

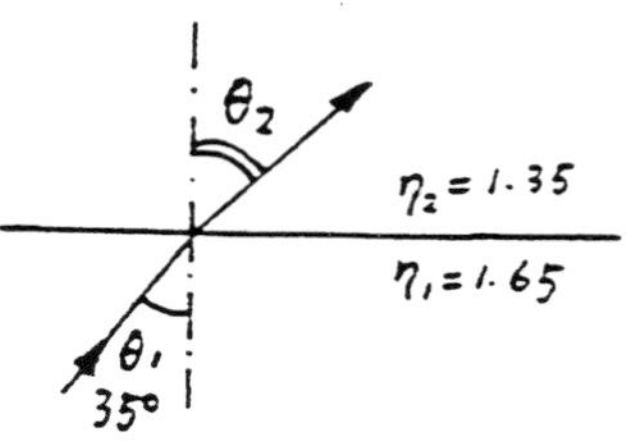

$$\sin\theta_2 = \frac{1.65}{1.35} \times \sin 35° = 0.7010$$

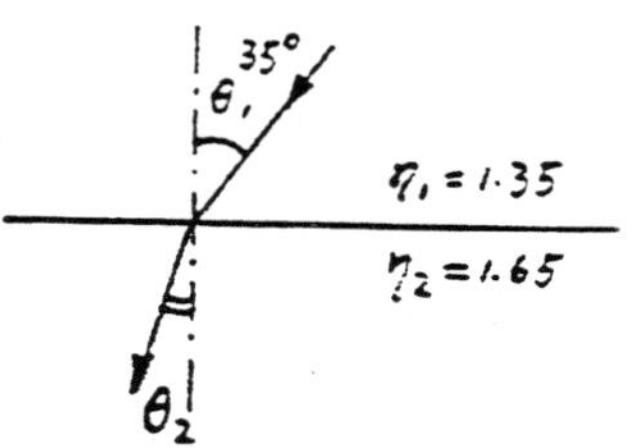

$\theta_2 = 44.51°$ or $44°30'37''$

b). $\sin\theta_2 = \frac{1.35}{1.65} \times \sin 35° = 0.4693$

$\theta_2 = 27.99°$ or $27°59'18''$

1.2 a) Refer to Example 1.5, because of the refraction, the apparent depth of each beaker is smaller than the real depth.

$$\frac{\sin G\hat{C}I}{\sin A\hat{C}B} = \frac{\eta}{\eta_{air}}$$

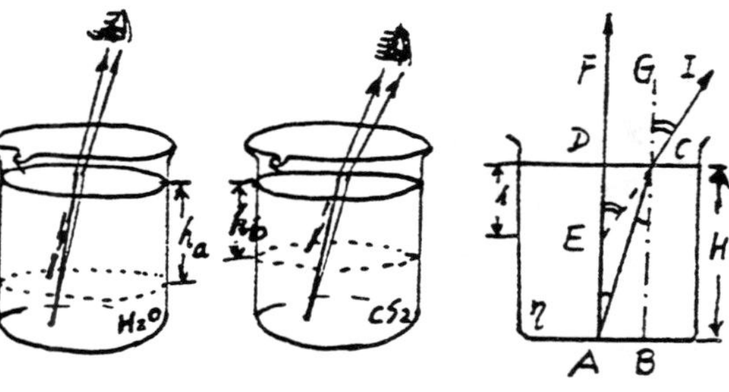

In fact these angles are very small,

$\sin\theta \simeq \theta \simeq tg\,\theta$

$$h = \frac{DC}{tg\,D\hat{E}C} = \frac{H\ tg\,D\hat{A}C}{tg\,D\hat{E}C} = \frac{H\ tg\,ACB}{tg\,GCI} = \frac{H}{\eta}$$

$\eta_{CS_2} > \eta_{H_2O}$, so $h_{CS_2} < h_{H_2O}$

b) $h_{CS_2} = \dfrac{H}{\eta_{CS_2}}$, $h_{H_2O} = \dfrac{H}{\eta_{H_2O}}$

$\therefore$ the ratio of apparent depth $h_{CS_2} : h_{H_2O} = \eta_{H_2O} : \eta_{CS_2}$

$= 1.33 : 1.63$ or $0.816 : 1$

1.3

From the same reasons in Problem 1.2, $h = \dfrac{d}{tg\,\gamma} \doteq \dfrac{d}{\gamma}$

$d = H_1\alpha + H_2\beta$

from Snell's law $\gamma = \eta_2\beta$, $\beta = \dfrac{\eta_1}{\eta_2}\alpha$

$$h = \frac{H_1\alpha + H_2\dfrac{\eta_1}{\eta_2}\alpha}{\eta_1\alpha} = \frac{H_1}{\eta_1} + \frac{H_2}{\eta_2}$$

$$= \frac{4}{1.33} + \frac{3}{1.36} = 3.00 + 2.21 = 5.21 \text{ cm}$$

1.4

Using the result in Problem 1.3, the apparent distance between the upper surface of the glass and the printed material $H' = \dfrac{H_1}{\eta_1} + \dfrac{H_2}{\eta_2} = \dfrac{5}{1} + \dfrac{4}{1.5} = 7.67$ cm

or in other words, the pinted material looks to be floating above the table, the "height" is $(5+4) - 7.67 = 1.33$ cm.

1.5 according to Snell's law, $\frac{\sin\beta}{\sin\alpha} = \frac{\eta_A}{\eta_G}$

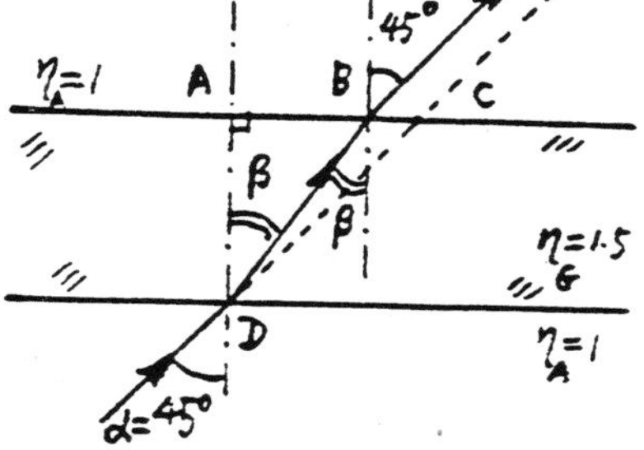

$$\sin\beta = \frac{1}{\eta_G}\sin 45^\circ = \frac{1}{1.5} \times \frac{\sqrt{2}}{2} = 0.4714$$

so $\beta = 28.13^\circ$ or $28^\circ 7' 32''$

The tranwverce displacement $BC = AC - AB$

$= AD(tg\,45^\circ - tg\,28^\circ 7' 32'') = 4 \times (1 - 0.5343) = 1.86$ cm

1.6 a). According to the Snell's law of refraction, Eq (1.9)

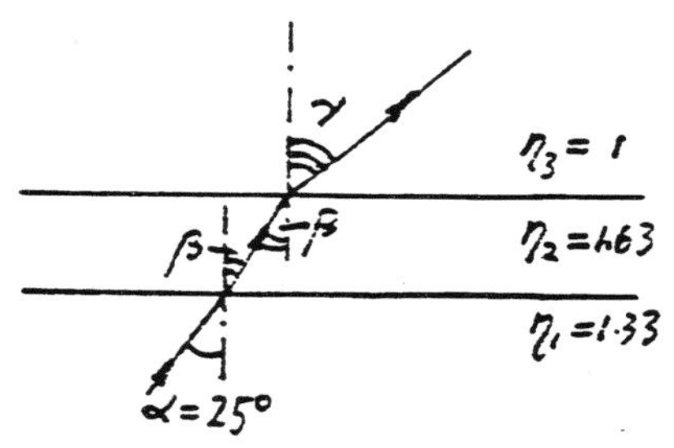

$$\frac{\sin\beta}{\sin\alpha} = \frac{\eta_1}{\eta_2}, \quad \sin\beta = \frac{\eta_1}{\eta_2}\sin\alpha$$

$$\sin\beta = \frac{1.33}{1.63} \times \sin 25^\circ = 0.3448$$

so the angle of refraction in the oil $\beta = 20.17^\circ$ or $20^\circ 10' 18''$.

b) $\frac{\sin\gamma}{\sin\beta} = \frac{\eta_2}{\eta_3}$, $\quad \sin\gamma = \eta_2 \times \sin\beta = 1.33 \times 0.3448$

$\sin\gamma = 0.4586 < 1$, in this case, the total inner reflection will not occur.

c) Refer to Eq (1.28) and Example 1.6 in this testbook, when β is critical angle of incidence $\gamma = 90^\circ$.

$$\frac{\sin\gamma}{\sin\beta} = \frac{\eta_2}{\eta_3}, \quad \sin\beta = \frac{1}{\eta_2} = 0.6135$$

$$\beta = \arcsin\frac{1}{\eta_2} = 37.84^\circ \quad \text{or} \quad 37^\circ 50' 34''$$

1.7

Let us consider the reflected wave first. See the following figure, the points P_1, P_2, ... on the boundary become the centers of secondary wave, according to Huygens' principle. The radii of these wavelets can be determined by Eq (1.11):

$$r_n = \frac{c\,\delta t_n}{\eta_1}$$

$$n = 1, 2, 3, \cdots$$

In the figure,

$r_0 = 0$

$r_1 = \lambda_1$

$r_2 = 2\lambda_1$

$r_3 = 3\lambda_1$

$r_4 = 4\lambda_1$,

draw an envelope of these wavelets.

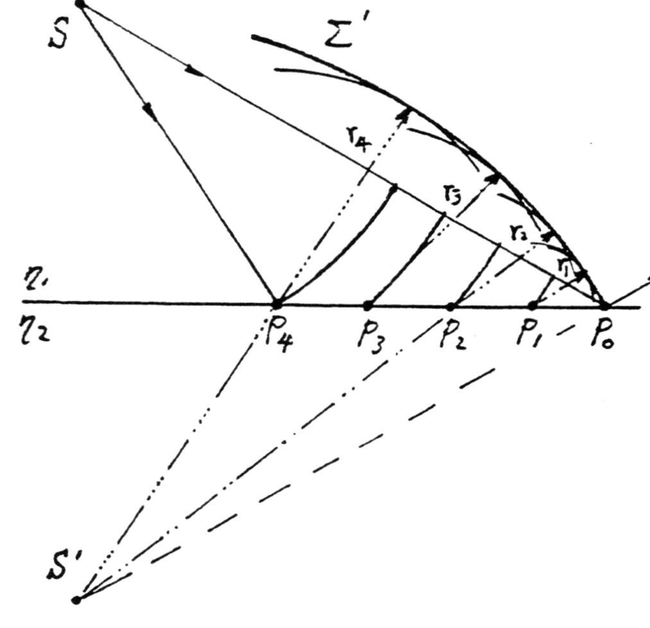

Σ' seems to be emitted by S', the image of the source S.

The wavelength of refracted light is λ_2, if $\eta_2 > \eta_1$, then $\lambda_2 < \lambda_1$. The radii of the refracted wavelets can be determined by Eq (1.14),

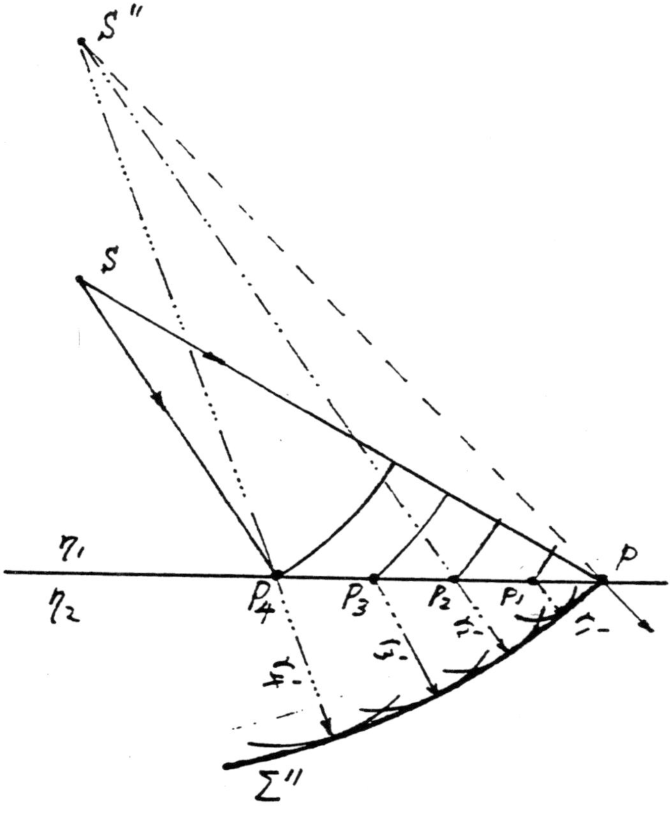

$$r'_n = \frac{c \Delta t_n}{\eta_2}$$

$n = 1, 2, 3, \cdots$

Here $r'_0 = 0$

$r'_1 = \lambda_2$

$r'_2 = 2\lambda_2$

$r'_3 = 3\lambda_2$

$r'_4 = 4\lambda_2$,

draw an envelope of refracted wavelets.

Σ'' seems to be emitted by S'', the image of the source S.

1.8 Refer to Example 1.5 or Problem 1.2

the apparent depth $h = \frac{H}{\eta}$ 1.5

So the real depth $H = \eta \cdot h = 1.33 \times 5 \text{ ft} = 6.65 \text{ ft}$.

1.9 According to Eq (1.28), the critical angle

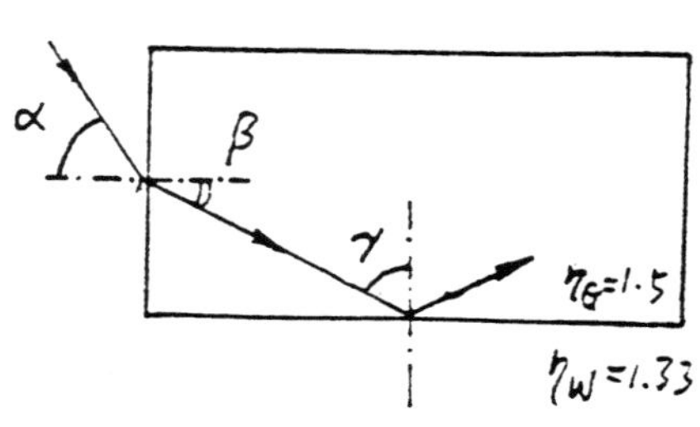

$$\theta_c = \arc \sin\left(\frac{\eta_{water}}{\eta_{Glass}}\right)$$

when $\gamma = \theta_c$, $\beta = \frac{\pi}{2} - \gamma$,

$$\sin\alpha_M = \frac{\eta_G}{\eta_W}\sin\beta = \frac{\eta_G}{\eta_W}\cos\gamma = \frac{\eta_G}{\eta_W}\sqrt{1-\left(\frac{\eta_W}{\eta_G}\right)^2}$$

$$= \sqrt{\left(\frac{\eta_G}{\eta_W}\right)^2 - 1} = \sqrt{\left(\frac{1.5}{1.33}\right)^2 - 1} = 0.5215$$

$\alpha_M = 31.43°$ or $31° 26' 2''$

If $\alpha > \alpha_M$, β will be larger and γ smaller than θ_c.

1.10 In this case, $\sqrt{\left(\frac{\eta_G}{\eta_{air}}\right)^2 - 1}$

$= \sqrt{2.25 \quad 1} = \sqrt{1.25} > 1$, the total inner reflection always occurs; no maximum limitation for α.

Even $\alpha = 90°$

$\beta = 41°$

$\gamma = 49° > \theta_c \ (=41°)$

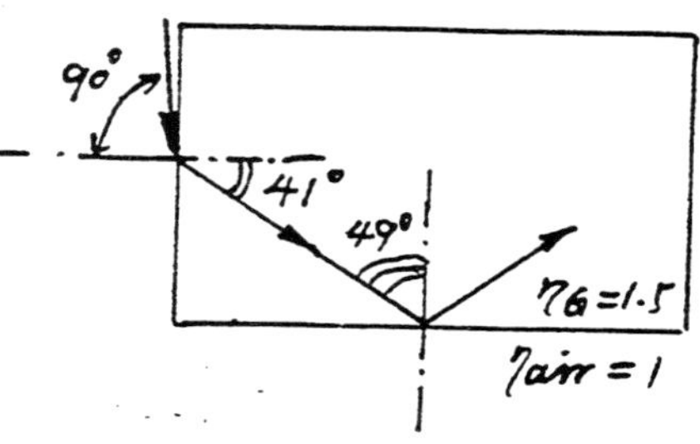

Smaller α means smaller β, even larger γ.

1.11 $\alpha = 45° = \theta_c$, refer to Eq (1.28)

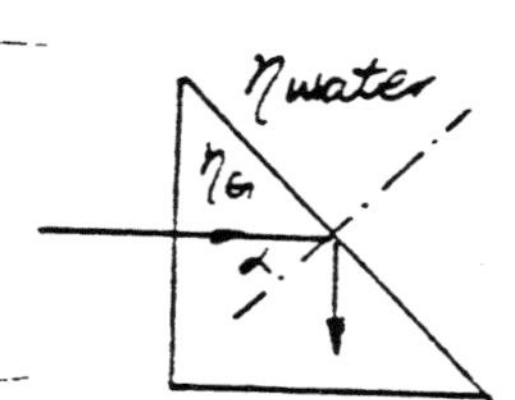

$$\sin\theta_c = \frac{\eta_r}{\eta_i},$$

$$\eta_G = \eta_i = \frac{\eta_r}{\sin\theta_c} = \frac{1.33}{\sin 45°} = 1.88$$

1.12 From the same reason in Problem 2.11,

$$\eta_L = \eta_r = \eta_i \cdot \sin\theta_c$$

$$= 1.6 \times \sin 60° = 1.386$$

η_L should be less than 1.386.

1.13 a) First, we determine the refractive index of crown glass : (FIG. 1.17)

$\lambda_1 = 700$ nm, $\eta_1 = 1.504$

$\lambda_2 = 400$ nm, $\eta_2 = 1.520$

Substitute η_1 and η_2 into Eq (1.31), the minimum deviation for different wavelengths :

$$\delta_1 \simeq (\eta_1 - 1)\phi = 0.504\,\phi,$$

$$\delta_2 \simeq (\eta_2 - 1)\phi = 0.520\,\phi,$$ where ϕ

is the apex angle of the prism.

b) From Fig. 1.17, we know

for blue light $\eta_f = 1.515$

for yellow light $\eta_d = 1.508$

for red light $\eta_c = 1.503$

According to Eq (2.36), the dispersive power is

$$W = \frac{\eta_f - \eta_c}{\eta_d - 1} = \frac{1.515 - 1.503}{1.508 - 1} = 0.0236$$

1.14

According to Fig 1.17, we can determine the refractive index of crown glass under different color lights:

$\eta_f = 1.515$

$\eta_d = 1.508$

$\eta_c = 1.503$

So the mean angular deviation

$\delta_d = (\eta_d - 1)\phi = 5.08°$ or $5° 4' 48''$

The dispersive power is the same as in Prob. 1.13, $W = \dfrac{\eta_f - \eta_c}{\eta_d - 1} = 0.0236$

1.15

a) After the first polarizer (P_1), the transmitted light is linearly polarized along the direction of P_1 and has an intensity of $I_1 = I_0/2$. Since polarizer P_2 is perpendicular to P_1, the transmitted light after P_2 is: $I_2 = I_1 \cos 90° = 0$. Therefore, the transmitted light intensity is zero.

b). Similarly, after the first polarizer P_1, we have $I_1 = I_0/2$. However after the inserted polarizer P_3 the transmitted light has an intensity of: $I_2 = I_1 \cos 45° = \sqrt{2}\, I_0/4$, which is linearly polarized in the direction of 45° with respect to that of polarizer P_2. Therefore, after the polarizer P_2, we have $I_3 = I_2 \cos 45° = I_0/4$.

1.16

a). For a nature light, the relative phase shift between two components is random in time, while for a circularly polarized light, the phase shift is fixed and equals $\pi/2$.

b). Again, for a partially linearly polarized light the phase shift is random while for an elliptically polarized light, it is fixed (but not necessarily equals $\pi/2$). Besides, the two perpendicular components are not equal in amplitude in both cases.

1.17

Since the critical angle is $\theta_c = \sin^{-1}(\eta_0/\eta_1) = 40°$, we can determine that the refractive index of the quarts is $\eta_1 = 1.56$. The Brewster's angle for the quartz is then

$$\theta_B = \tan^{-1}(\eta_0/\eta_1) = 57.3°$$

1.18 Intensity of plane wave:

$$I = \frac{1}{2}\sqrt{\frac{\varepsilon}{\mu}} \cdot (E_0)^2$$

Energy per unit area of incident and reflected beam

$$W_1 = I_1 \cos\theta_1 \cdot (E_1)^2 = \frac{1}{2}\sqrt{\frac{\varepsilon_1}{\mu_1}} \cos\theta_1 \, (E_1)^2$$

$$W_2 = I_2 \cos\theta_1 \cdot (E_2)^2 = \frac{1}{2}\sqrt{\frac{\varepsilon_2}{\mu_2}} \cos\theta_2 \cdot (E_2)^2$$

$$T = \frac{W_2}{W_1} = \frac{\frac{1}{2}\sqrt{\frac{\varepsilon_2}{\mu_2}} \cos\theta_2 (E_2)^2}{\frac{1}{2}\sqrt{\frac{\varepsilon_1}{\mu_1}} \cos\theta_1 (E_1)^2} = \frac{\eta_2 \cos\theta_2}{\eta_1 \cos\theta_1} \cdot \left(\frac{E_2}{E_1}\right)^2$$

$$\therefore \quad T_s = \frac{\eta_2 \cos\theta_2}{\eta_1 \cos\theta_1} \cdot \frac{4\sin^2\theta_2 \cos^2\theta_1}{\sin^2(\theta_1 + \theta_2)}$$

$$T_p = \frac{\eta_2 \cos\theta_2}{\eta_1 \cos\theta_1} \cdot \frac{4\sin^2\theta_2 \cos^2\theta_1}{\sin^2(\theta_1 + \theta_2)\cos^2(\theta_1 - \theta_2)}$$

1.19

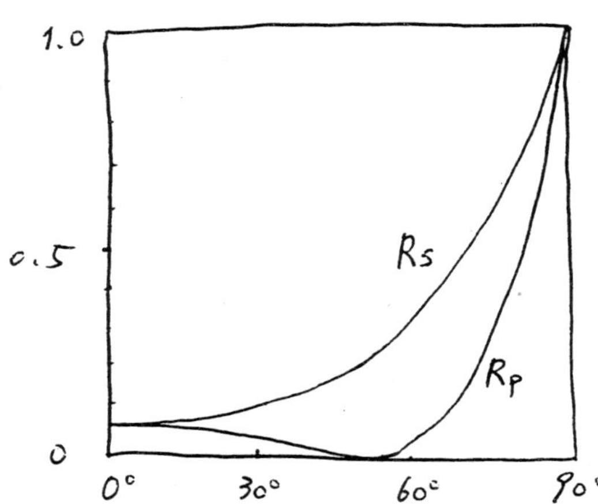

1.20

$$R = \left(\frac{\eta_2 - \eta_1}{\eta_2 + \eta_1}\right)^2$$

$\eta_1 = 1.0$, $\therefore$ $R = \left(\frac{\eta_2 - 1}{\eta_2 + 1}\right)^2$

a) $R = 4.34\%$

(b). $R = 4.61\%$

(c). $R = 17.32\%$

(d). $R = 24.32\%$

Chapter 2

2.1 a)

Substitute the given data into

Eq (2.4): $\frac{\eta_1}{d_1} + \frac{\eta_2}{d_2} = \frac{\eta_2 - \eta_1}{R}$

$\eta_1 = 1.5$ $\eta_2 = 1$ C

$\frac{1.5}{100} + \frac{1}{d_2} = \frac{1 - 1.5}{20}$ so $d_2 = -\frac{200}{7}$ mm

Using Eq (2.12), the lateral magnification

$$m = -\frac{\eta_1 d_2}{\eta_2 d_1} = -\frac{1.5 \times (-200/7)}{1 \times 100} = \frac{3}{7} \simeq 0.43$$

b) We get an upright reduced virtual image at the same side of the object (28.7 mm away from the vertex), in stead of an inverted real image in Example 2.1.

2.2

a) Using Eq (2.5) and (2.6), we have

$$f_2 = \frac{\eta_2}{\eta_2 - \eta_1} R = \frac{1.5}{1.5 - 1} \times 20 = 60 \text{ mm}$$

$$f_1 = \frac{\eta_1}{\eta_2 \ \eta_1} R = \frac{1}{1.5 - 1} \times 20 = 40 \text{ mm}$$

b)

$\eta_1 = 1$ $\eta_2 = 1.5$ F_2 f_2

F_1 $\eta_1 = 1$ $\eta_2 = 1.5$ f_1

F_2 is located in medium 2, so the given spherical convex surface is a converging optical element. F_1 is in medium 1.

2.3 $f_2 = \frac{\eta_2}{\eta_2 - \eta_1} R = \frac{1}{1-1.5} \times 20 = -40$ mm

$f_1 = \frac{\eta_1}{\eta_2 - \eta_1} R = \frac{1.5}{1-1.5} \times 20 = -60$ mm

$\eta_1 = 1.5$ / $\eta_2 = 1$

F_2

$\eta_1 = 1.5$ / $\eta_2 = 1$

F_1

In this case, both F_2 and F_1 are virtual rather than real focal points. Here, the spherical surface is concave, serves as diverging optical element.

2.4 Use Eq (2.4) once again:

$$\frac{\eta_1}{d_1} + \frac{\eta_2}{d_2} = \frac{\eta_2 - \eta_1}{R}$$

$\eta_1 = 1.5$ $\eta_2 = 1$

that is $\frac{1.5}{20} + \frac{1}{d_2} = \frac{1-1.5}{20}$ $\quad d_2 = -10$ mm

A diverging optical element gives a reduced virtual image.

2.5 Substitute the given data into Eq (2.4)

$$\frac{1.33}{15} - \frac{1}{d_2} = \frac{1-1.33}{(-15)} \qquad d_2 = -15 \text{ cm}$$

the magnification $m = -\frac{\eta_1 d_2}{\eta_2 d_1}$

so $m = -\frac{1.33 \times (-15)}{1 \times 15}$

$= 1.33$

There is an enlarged virtual image at the center of the spherical fish bowl.

2.6 For reflecting mirror we may substitute $\eta_2 = -1$ into the lens maker's equation (2.4)

$$\frac{\eta_1}{d_1} + \frac{\eta_2}{d_2} = \frac{\eta_2 - \eta_1}{-R}$$

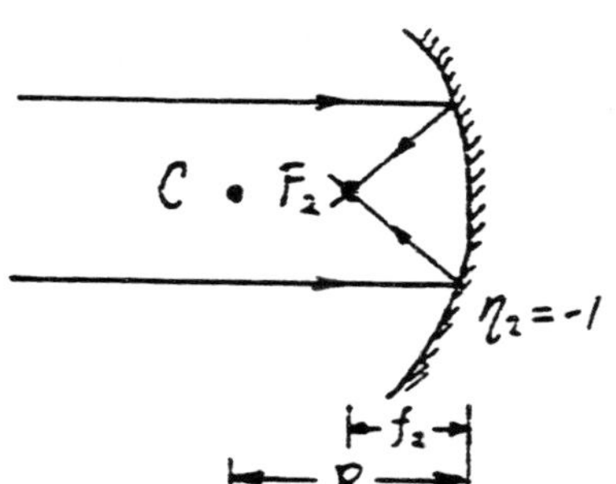

$\eta_1 = 1$, $\eta_2 = -1$, $d_1 = \infty$

so $f_2 = d_2 = -\frac{R}{2}$

2.7 In this problem, $d_1 = 2 \times f = R$

a) $\frac{\eta_1}{d_1} + \frac{\eta_2}{d_2} = \frac{\eta_2 - \eta_1}{-R}$,

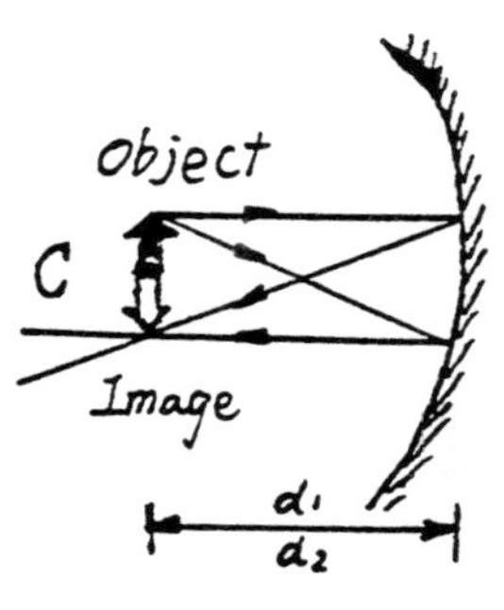

$\frac{1}{R} + \frac{-1}{d_2} = \frac{-1-1}{-R}$, so $d_2 = -R$

b) the lateral magnification

$$m = -\frac{\eta_1 d_2}{\eta_2 d_1} = -\frac{1 \times (-R)}{(-1) \times R} = -1$$

So this is an inverted equal-sized real image.

2.8

We can either subsitute a relative index of refraction $\eta_r = \frac{\eta}{\eta'}$ into Eq (2.20):

$$f = \frac{1}{(\frac{\eta}{\eta'} - 1)(\frac{1}{R_1} + \frac{1}{R_2})} = \frac{\eta' R_1 R_2}{(\eta - \eta')(R_1 + R_2)}$$

or let $\eta_1 = \eta_2 = \eta'$ (see the answer to next problem). Both methods yield the same result.

2.9

By using Eq (2.4) repeatedly, we obtain

$$\frac{\eta_1}{\infty} + \frac{\eta}{d} = \frac{\eta - \eta_1}{R_1}$$

so $\frac{\eta}{d} = \frac{\eta - \eta_1}{R_1}$

for the first surface. Then for the second surface:

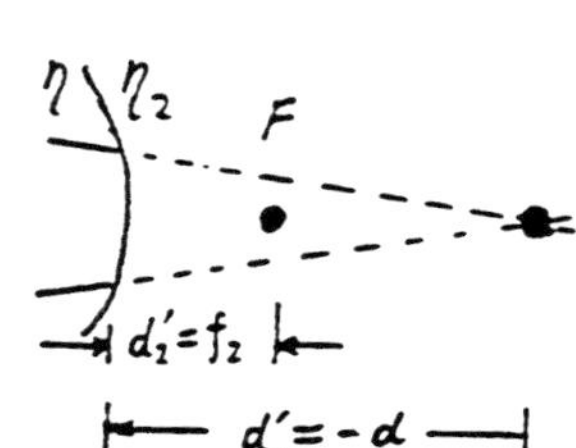

$$\frac{\eta}{-d} + \frac{\eta_2}{f_2} = \frac{\eta_2 - \eta}{-R_2}$$

(Note. ① for thin lens, $d_1' = -d_2$.
② for concave surface, the radius is negative.)

thus $\frac{\eta_2}{f_2} = \frac{\eta - \eta_2}{R_2} + \frac{\eta - \eta_1}{R_1}$

$$f_2 = \frac{\eta_2}{\frac{\eta - \eta_1}{R_1} + \frac{\eta - \eta_2}{R_2}}$$

To calculate the first focal length, we consider the second surface first. assume $d_2' \infty$

$$\frac{\eta}{d_1'} + \frac{\eta_2}{\infty} = \frac{\eta_2 - \eta}{-R_2}$$

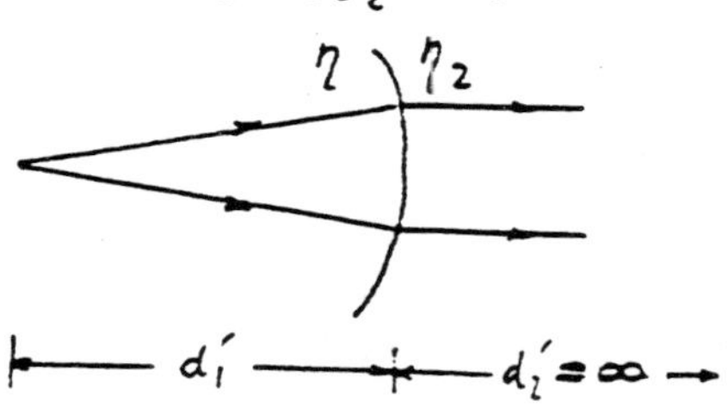

$$\frac{\eta}{d_1'} = \frac{\eta - \eta_2}{R_2}$$

then for the first surface,

$$\frac{\eta_1}{f_1} + \frac{\eta}{-d_1'} = \frac{\eta - \eta_1}{R_1}$$

$$\frac{\eta_1}{f_1} = \frac{\eta}{d_1'} + \frac{\eta - \eta_1}{R_1}$$

$$= \frac{\eta - \eta_2}{R_2} + \frac{\eta - \eta_1}{R_1}$$

so $$f_1 = \frac{\eta_1}{\frac{\eta - \eta_1}{R_1} + \frac{\eta - \eta_2}{R_2}}.$$

2.10

Just let $\eta_1 = \eta_{air} = 1$, substitute this into the results of Problem 2.9, we have

$$f_1 = \frac{1}{\frac{\eta - 1}{R_1} + \frac{\eta - \eta_2}{R_2}},$$

and

$$f_2 = \frac{\eta_2}{\frac{\eta - 1}{R_1} + \frac{\eta - \eta_2}{R_2}}.$$

2.11 Use the result of Problem 2.9, the first and second focal lengths are

$$f_1 = \frac{\eta_1}{\frac{\eta-\eta_1}{R_1} + \frac{\eta-\eta_2}{R_2}} = \frac{1}{\frac{1.5-1}{10} + \frac{1.5-1.33}{15}} = \frac{1}{0.0613} = 16.3 \text{ cm},$$

$$f_2 = \frac{\eta_2}{\frac{\eta-\eta_1}{R_1} + \frac{\eta-\eta_2}{R_2}} = \frac{1.33}{0.0613} = 21.7 \text{ cm, respectively.}$$

2.12 To find the relation between the object and the image, let us observe two pairs of triangles in the following illustration:

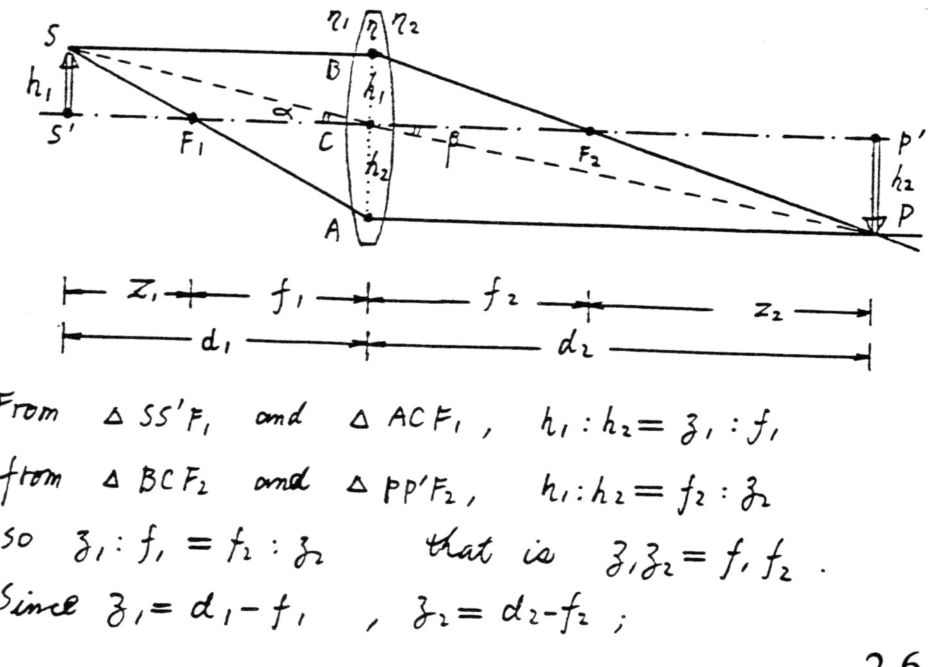

From $\Delta\, SS'F_1$ and $\Delta\, ACF_1$, $h_1 : h_2 = z_1 : f_1$

from $\Delta\, BCF_2$ and $\Delta\, PP'F_2$, $h_1 : h_2 = f_2 : z_2$

so $z_1 : f_1 = f_2 : z_2$ that is $z_1 z_2 = f_1 f_2$.

Since $z_1 = d_1 - f_1$, $z_2 = d_2 - f_2$;

$$(d_1 - f_1)(d_2 - f_2) = f_1 f_2 \qquad \text{that is}$$

$$\frac{f_1}{d_1} + \frac{f_2}{d_2} = 1$$

$$d_2 = \frac{f_2 d_1}{d_1 - f_1}, \quad \text{here } d_2 = \frac{21.7}{1 - \frac{1}{1.5}} = 21.7 \times 3$$

$$= 65.1 \text{ cm}$$

According to the Snell's law of refration

$\frac{\sin\alpha}{\sin\beta} = \frac{\eta_1}{\eta_2}$, when α and β are small.

$\sin\alpha = \frac{h_1}{d_1}$, $\sin\beta = \frac{h_2}{d_2}$, so the lateral magnification $m = \frac{h_2}{h_1} = -\frac{\eta_1 d_2}{\eta_2 d_1}$ (for thin lens)

here $m = -\frac{65.1 \times 1}{16.3 \times 1.5 \times 1.33} = -2.0$

m may also be calculated according to

$m = \frac{f_1}{f_1 - d_1}$ (see the illustration), yielding the same result.

2.13

Here $\eta_1 = 1.33$, $\eta_2 = 1$. $f_1 = 21.7$ cm. $f_2 = 16.3$ cm.

$d_1 = 1.5 f_1$, so

$$d_2 = \frac{f_2}{1 - f_1/d_1} = \frac{16.3}{1 - \frac{1}{1.5}} = 16.3 \times 3 = 48.9 \text{ cm}$$

$$m = -\frac{1.33 \times 48.9}{1 \times 21.7 \times 1.5} = -2.0$$

2.14

Substitute the given data into Eq.(2.20):

$$f = \frac{1}{(1.5-1)(\frac{1}{15}+\frac{1}{15})} = \frac{15}{0.5\times 2} = 15 \text{ cm}$$

then turn to Eq.(2.25)

$$\frac{1}{f/2} + \frac{1}{d_2} = \frac{1}{f} \qquad \therefore \; d_2 = -f = -15 \text{ cm}$$

the lateral magnification (see Eq. 2.24)

$$m = -\frac{d_2}{d_1} = -\frac{-f}{f/2} = 2$$

2.15

The thickness variation here

$t = R - \sqrt{R^2 - \rho^2}$,

where $\rho^2 = x^2 + y^2$.

Under paraxial approximation,

$$t \simeq \frac{\rho^2}{2R}$$

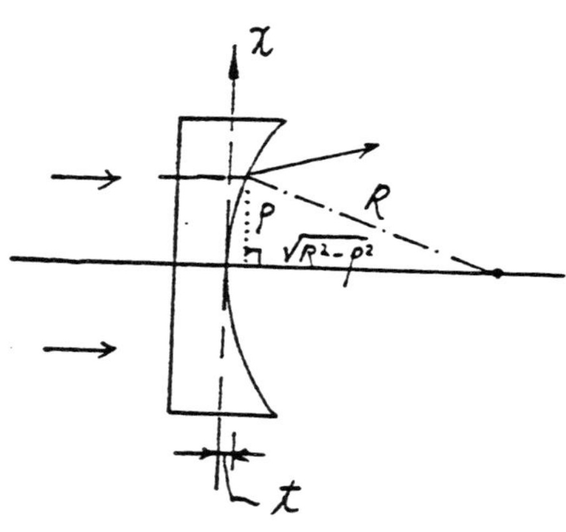

So the phase transform $t(x,y) = \exp[ik(n-1)t]$

$$= e^{i\frac{k(n-1)}{2R}(x^2+y^2)}$$

This means its divergent effect. In other words, a

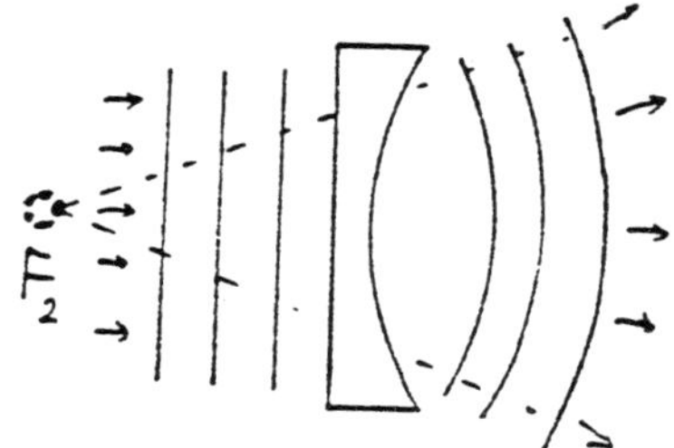

normally incident plane wave will be converted into a divergent wave with its center at the virtual focal point F_2.

2.16

a) According to Eq (2.37) and Example 2.7, the phase transform of the first lens is

$$T_1 = e^{-i\frac{k}{2f_1}\rho^2}, \quad \text{where } f_1 = \frac{R_{11}R_{12}}{(n_1-1)(R_{11}+R_{12})}.$$

and for the second lens:

$$T_2 = e^{-i\frac{k}{2f_2}\rho^2}, \quad \text{where } f_2 = -\frac{R_{12}R_{22}}{(n_2-1)(R_{12}+R_{22})}.$$

so the phase transform of the doublet lens

$$T = T_1 \cdot T_2 = C\, e^{-i\frac{k}{2}(x^2+y^2)(\frac{1}{f_1}+\frac{1}{f_2})}$$

b) Compare this equation and (3.37), the resultant focal length is $f = \frac{f_1 f_2}{f_1+f_2}$.

c) $U_i(x,y) \rightarrow \otimes \rightarrow \otimes \rightarrow U_o(x,y)$

with T_1 feeding the first $\otimes$ and T_2 feeding the second $\otimes$.

$$T_1 = c\, e^{-ik\frac{(x^2+y^2)}{2}(\eta_1-1)\left(\frac{1}{R_{11}}+\frac{1}{R_{12}}\right)}$$

$$T_2 = c\, e^{ik\frac{(x^2+y^2)}{2}(\eta_2-1)\left(\frac{1}{R_{12}}+\frac{1}{R_{22}}\right)}$$

2.17

We use the Gaussian lens equation (2.25), as if the doublet lens were a single lens with a focal length f.

$$d_2 = \frac{f d_1}{f - d_1} = \frac{f \times 1.2f}{f - 1.2f} = -6f .$$

The lateral magnification [see Eq (2.24)]

$$M = -\frac{d_2}{d_1} = -\frac{-6f}{1.2f} = 5$$

2.18

a) $U_i(x,y) \rightarrow \otimes \rightarrow \otimes \rightarrow U_o(x,y)$

with T_1 feeding the first $\otimes$ and T_2 feeding the second $\otimes$.

b) According to Eq (2.36), focal length of each single lens :

$$f_1 = \frac{1}{(\eta_1-1)\left(\frac{1}{R_{11}}+\frac{1}{R_{12}}\right)} = \frac{R_{11}}{\eta_1-1} ,$$

where $R_{12} = \infty$.

$$f_2 = \frac{1}{(\eta_2 - 1)\left(\frac{1}{R_{12}} + \frac{1}{-R_{22}}\right)} = -\frac{R_{22}}{\eta_2 - 1}$$

For contact thin lenses combination, the over-all focal length is

$$f = \frac{f_1 f_2}{f_1 + f_2} \qquad \text{(Refer to Example 2.7)}$$

$$f = \frac{\frac{R_{11}}{\eta_1 - 1} \cdot \frac{-R_{22}}{\eta_2 - 1}}{\frac{R_{11}}{\eta_1 - 1} - \frac{R_{22}}{\eta_2 - 1}} = \frac{R_{11} R_{22}}{R_{22}(\eta_1 - 1) - R_{11}(\eta_2 - 1)}$$

2.19

From Fig 1.17, the refractive indices of flint grass for different colors are approximately

$\eta_v = 1.658$, for violet, $\lambda_v = 400$ nm

$\eta_r = 1.608$, for red, $\lambda_r = 700$ nm

a) Substitute $R = 20$ cm into Eq.(2.36)

$$f = \frac{R_1 R_2}{(\eta - 1)(R_1 + R_2)} = \frac{R}{2(\eta - 1)}$$

so $$f_v = \frac{20}{2 \times (1.658 - 1)} = 15.20 \text{ cm}$$

$$f_r = \frac{20}{2 \times (1.608 - 1)} = 16.45 \text{ cm}$$

There is a smeared focal line in stead of focal point behind the lens.

b) According to the Gaussian lens equation (Eq 2.25), $d_2 = \dfrac{d_1 f}{d_1 - f}$.

For violet illumination,

$$d_{2v} = \frac{d_1 f_v}{d_1 - f_v} = \frac{30 \times 15.20}{30 - 15.20} = 30.81 \text{ cm}.$$

For red illumination,

$$d_{2r} = \frac{d_1 f_r}{d_1 - f_r} = \frac{30 \times 16.45}{30 - 16.45} = 36.42 \text{ cm}.$$

The lateral magnification $m = -\dfrac{d_2}{d_1}$,

so $m_v = -\dfrac{30.81}{30} = -1.03$.

$m_r = -\dfrac{36.42}{30} = -1.21$. (See Fig 2.18.)

2.20

According to Eq. (2.40)

$h_{1\,min} = \dfrac{1.22\ \lambda d_1}{D}$, so $D = \dfrac{1.22\ \lambda d_1}{h_{1\,min}}$

a) Under red light illumination, the required aperture

$$D_r = \frac{1.22\ \lambda_r d_1}{h_{1\,min}} = \frac{1.22 \times 650 \times 10^{-6} \times 500}{0.001} = 396.5 \text{ mm}$$

b) Under violet light illumination.

$$D_V = \frac{1.22\,\lambda_V\, d_1}{h_{1\,min}} = \frac{1.22 \times 400 \times 10^{-6} \times 500}{0.001} = 244 \text{ mm}$$

We see, for shorter wavelength, the size of the imaging lens can be reduced

2.21

From Eq (2.39), $\Psi_{min} = \frac{1.22\,\lambda}{D}$, the minimum angular separation in this case

$$\Psi_{min} = \frac{1.22 \times 550 \times 10^{-6} \text{ mm}}{10 \text{ mm}} = 6.71 \times 10^{-5} \text{ (Radian).}$$

The minimum resolvable separation of the image $h_{2\,min} = \Psi_{min}\, d_2$, where d_2 is the distance between the lens and the image

$d_2 = \frac{d_1 f}{d_1 - f}$, f is the focal length.

2.22

a) The minimum angular separation here is

$$\Psi_{1\,min} = \frac{1.22\,\lambda_1}{D} = \frac{1.22\,\lambda}{D\,\eta_1} = \frac{\Psi_{min}}{\eta_1} = 4.79 \times 10^{-5}$$

b) The oil immersion makes a smaller minimum angular resolvable separation possible.

2.23

a) Since the object space is filled with air, the same result as appeared in Prob. 2.21 will be obtained,

$$\psi_{1\,min} = \frac{1.22\lambda}{D} = 6.71 \times 10^{-5} \quad \text{(Radian)}$$

b) In this case, the numerical aperture

$$N.A. = \eta_1 \sin\theta_1 = 1 \times \sin(\operatorname{arc\,tg} \frac{0.5}{100}) = 0.05$$

c) In problem 2.22, $\psi_{1\,min} = 4.79 \times 10^{-5}$, and $N.A = \eta_1 \sin\theta_1 = 1.4 \times 0.05 = 0.07$. So the presence of oil in the object space means larger N.A. and smaller resovable separation, but the presence of oil in the image space does not make sence.

2.24

a) According to Eq (2.39), the minimum angular resovable separation (in object space)

$$\psi_{1\,min} = \frac{1.22\lambda_1}{D} = \frac{1.22\lambda}{\eta_1 D}, \quad h_{1\,min} = \psi_{1\,min}\, d_1$$

In image space, $\psi_{2\,min} = \frac{\eta_1}{\eta_2}\psi_{1\,min} = \frac{1.22\lambda}{\eta_2 D}$

and $h_{2\,min} = \psi_{2\,min}\, d_2$. (Snell's law).

b) $N.A. = \eta_1 \sin\theta_2$

Chapter 3

3.1 Intensity received by retina is proportional to the diameter of iris, i.e.,

$$I \propto A = \pi\left(\frac{d}{2}\right)^2 .$$

When $d_1 = 2$ mm, $d_2 = 6$ mm, we have

$$\frac{I_2}{I_1} \doteq \frac{\pi \times \left(\frac{6}{2}\right)^2}{\pi \times \left(\frac{2}{2}\right)^2} = 9 .$$

Thus, the intensity received by retina increases by nine times.

3.2 From imaging equation, we have

$$\frac{1}{f} = \frac{1}{\infty} + \frac{1}{d}$$

$$d = f = 2 \text{ m}.$$

The far point is two meters from the eye.

3.3 A positive lens is needed. The focal length is given by:

$$\frac{1}{f} = \frac{1}{25} + \frac{1}{(-200)}$$

$$f = 28.57 \text{ cm}.$$

3.4 Substitute $f = 50$ mm and $d_1 = 5$ m into imaging equation, we have

$$\frac{1}{f} = \frac{1}{d_1} + \frac{1}{d_2}$$

$$d_2 = 50.5 \quad mm.$$

The magnification is

$$\beta = \frac{d_2}{d_1} \approx 0.01^{\times}.$$

$$\frac{25}{0.01} = 2500 \quad mm,$$

$$\frac{36}{0.01} = 3600 \quad mm.$$

Thus, the field of view is $2.5 \times 3.6 \ m^2$.

3.5 From Eqs. (3.1) and (3.2), we know

$$\frac{t_2}{t_1} = \left(\frac{d_1}{d_2}\right)^2$$

when $d_1 = \frac{f}{16}$, $d_2 = \frac{f}{5.6}$, we have

$$\frac{t_2}{t_1} = \left(\frac{5.6}{16}\right)^2 = 0.1225.$$

That is : $t_2 \approx \frac{1}{8} t_1$

3.6 a). $D = \gamma \log E - D_0$

From fig. 3.16, we get

$E = 1000 \longrightarrow D = 1.2$

$E = 1800 \longrightarrow D = 2.3$

so $\begin{cases} 1.2 = r \log 1000 - D_o \\ 2.3 = r \log 1800 - D_o \end{cases} \Rightarrow \begin{cases} r = 4.31 \\ D_o = 11.73 \end{cases}$

b). $D = 4.31 \log E - 11.73$

The linear region : $1.2 \leq D \leq 3.0$

It corresponds to : $1000 \leq E \leq 2616$

3.7 The angular magnification is

$$M = -\frac{f_o}{f_e} = -\frac{262.5}{25} = -10.5^{\times}$$

3.8 $f_o = 16D = 16 \times 25 = 400$ mm.

$$M = -\frac{f_o}{f_e} = -\frac{400}{20} = -20^{\times}$$

3.9 He can replace the positive (convex) eyepiece by a negative (concave) one.

3.10 When a student looks through a telescope backward, everything looks smaller and appears like far away. This is because of the small magnification (smaller than one).

3.11 (a). $f_o + x' + f_e = 180$

$x' = 180 - f_o - f_e = 180 - 8.2 - 5.2 = 166.6$ mm

Image distance is

$d_2 = f_o + x' = 174.8$ mm.

Object distance is given by

$$\frac{1}{d_1} + \frac{1}{d_2} = \frac{1}{f_o}$$

$d_1 = 8.6$ mm.

(b). $m_1 = -\frac{x'}{f_o} = -\frac{166.6}{8.2} = 20.32^{\times}$

(c). $M = -\frac{x'}{f_o} \cdot \frac{250}{f_e} = -\frac{166.6 \times 250}{8.2 \times 5.2} = 977^{\times}$

3.12 (a). $M = m_1 \frac{250}{f_e} = -20 \times \frac{250}{25} = 200^{\times}$

(b). From conditions given in the problem, we have

$$\begin{cases} \beta = \frac{d_2}{d_1} = 20^{\times} \\ \frac{1}{d_1} + \frac{1}{d_2} = \frac{1}{f_o} \end{cases}$$

Solve above equations, we get

$d_2 = 21 f_o$

The distance from the objective to the eyepiece is

$\Delta = d_2 + f_e = 21 f_o + f_e$

3.13 From the given conditions, we have

$$\begin{cases} x' + f_o + f_e = 180 \text{ mm} \\ M = -\dfrac{x'}{f_o} \cdot \dfrac{250 \text{ mm}}{f_e} = -300^{\times} \\ f_o = 5 \text{ mm} \end{cases}$$

Solving above equations, we get

$$f_e = 25 \text{ mm}.$$

3.14 (a). $\dfrac{1}{d_1} = \dfrac{1}{f} - \dfrac{1}{d_2} = \dfrac{1}{75 \text{ mm}} - \dfrac{1}{12 \text{ m}}$

$d_1 = 75.47 \text{ mm}$

$M = \dfrac{d_2}{d_1} = \dfrac{12 \text{ m}}{75.47 \text{ mm}} \approx 160^{\times}$

(b). Before moving the projector, the object distance is

$\dfrac{1}{d_1'} = \dfrac{1}{f} - \dfrac{1}{d_2'} = \dfrac{1}{75 \text{ mm}} - \dfrac{1}{5 \text{ m}}$

$d_1' = 76.14 \text{ mm}.$

$\Delta = d_1' - d_1 = 0.67 \text{ mm}.$

3.15 The solid angle of the Fresnel lens to the light source is about

$$\int_0^{2\pi}\int_0^{30^\circ} \sin\theta \cdot d\theta \cdot d\varphi = 0.268\,\pi .$$

Thus: $\eta = \dfrac{0.268\,\pi}{4\pi} = 6.7\%.$

Chapter 4

4.1 Literature Search. No solution is provided.

4.2 The relationship between energy difference and wavelength can be expressed as

$$\lambda = \frac{1.24}{\Delta E},$$

where λ is measured in μm and ΔE in eV. When $\Delta E = 0.05$ eV, the cut-off wavelength is

$$\lambda = \frac{1.24}{0.05} = 24.8 \ \mu m.$$

4.3 From Eq. (4.2)

$$i = \eta \frac{e}{h\nu}\left(\frac{\tau_0}{\tau_d}\right) I = 10\% \cdot \frac{1.6\times10^{-19}}{6.626\times10^{-34}\cdot\nu}\left(\frac{10^{-5}}{10^{-7}}\right) I$$

$$= 2.415\times10^{15}\left(\frac{I}{\nu}\right),$$

$$V_b = i\,(R_t + R),$$

$$R_t = \frac{V_b}{i} - R = 4.14\times10^{-16}\cdot\frac{\nu\cdot V_b}{I} - R$$

4.4 $l_d = 0.5\ \mu m, \quad V_d = 10^7\ cm/sec$

$$t_d = \frac{l_d}{V_d} = \frac{0.5\ \mu m}{10^7\ cm/sec} = 0.5 \times 10^{-11}\ sec.$$

4.5 $E_g < h\nu = h \cdot \frac{c}{\lambda}$

$$= 8.6 \times 10^{-34} \times \frac{3 \times 10^8}{1.06 \times 10^{-6}}$$

$$= 2.434 \times 10^{-19}\ J$$

$$= 1.17\ eV$$

$\therefore \quad E_{g\,max} = 1.17\ eV.$

4.6 See text.

4.7 $l_{in} = 2.5\ \mu m, \quad \tau = 0.05\ nsec.$

$$V_d = \frac{l_{in}}{\tau} = \frac{2.5 \times 10^{-4}\ cm}{0.05 \times 10^{-9}\ sec}$$

$$= 5 \times 10^6\ cm/sec.$$

4.8

$$dM = \frac{1000\, V^2 \cdot e^{-\frac{V}{h}}}{(1 + 10e^{-\frac{V}{h}})^2\, h^3} \cdot dh \ \underline{\underline{let}}\ 5\%$$

$$dh \leq 5\% \cdot \frac{(1 + 10e^{-V_o/h_o})^2\, h_o^3}{1000\, V_o^2\, e^{-\frac{V_o}{h_o}}}$$

where V_o and h_o are nominal voltage and thickness.

4.9 $\lambda = 0.3\ \mu m$

$$W = h\nu = h\frac{c}{\lambda} = 4.13\ eV$$

4.10 $G = 4^4 \cdot 5^6 = 4 \times 10^6$

4.11 $1\ W = 634\ lumin$

$$\frac{1\ lumin}{1\ m^2} = 1\ lux$$

$\therefore$ $10^{-9}\ W/cm^2 = 10^{-4}\ W/m^2 = 0.634\ lux$

$> 0.1\ lux$

$\therefore$ He can use the CCD to videotape the moon.

4.12 Assume the greyscale of the CCD is 8 bits. Then the data rate is

$$r = 512 \times 512 \times 30 \times 8$$

$$= 62.9 \times 10^6\ bit/sec.$$

4.13 Consider two electrodes, as shown in the figure. Electrons emit from A to B.

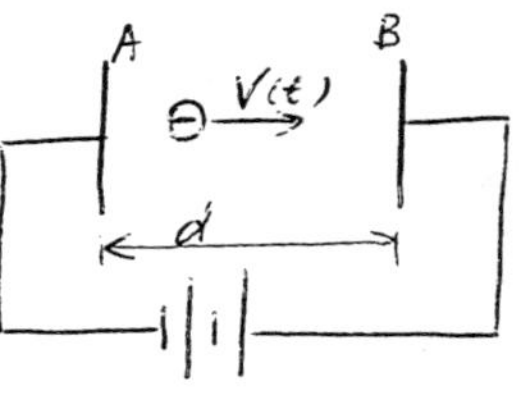

The average rate is $\bar{N}$, then

$$\bar{I} = \frac{\bar{N}}{e}$$

$$i_e(t) = \frac{e \cdot N(t)}{d}$$

$$F(\omega) = \frac{e}{2\pi d} \int_0^{t_{ab}} N(t)\, e^{-i\omega t}\, dt$$

If: $\omega \cdot t_{ab} \ll 1$, i.e., $i_e(t) \simeq c \cdot \delta(t)$

Then:

$$F(\omega) = \frac{e}{2\pi d} \int_0^{t_{ab}} \frac{dx}{dt} \cdot dt = \frac{e}{2\pi}$$

Power density:

$$S(\omega) = 8\pi^2 \bar{N} \cdot |F(\omega)|^2$$

$$= 8\pi^2 \bar{N} \cdot \left(\frac{e}{2\pi}\right)^2 = 2e^2\bar{N} = 2e\bar{I}$$

$\omega = 2\pi\nu$.

$$\therefore \quad \overline{i^2}(\nu) = S(\nu) \cdot \Delta\nu = 2e\bar{I} \cdot \Delta\nu$$

4.14 From Eq. (5.16)

$$\Delta\nu = \frac{P_{min} \cdot \eta}{h\nu_c} = \frac{0.08 \times 10^{-18} \times 1.06 \times 10^{-6}}{6.26 \times 10^{-14} \cdot 3 \times 10^{+8}}$$

$$= 0.5 \text{ Hz}.$$

4.15: (a) $\lambda = 488\ nm$, $h\nu = 2.54\ eV$

(b) $\lambda = 514.5\ nm$, $h\nu = 2.41\ eV$

(c) $\lambda = 930\ nm$, $h\nu = 1.33\ eV$

(d) $\lambda = 1.06\ \mu m$, $h\nu = 1.17\ eV$

(e) $\lambda = 10.6\ \mu m$, $h\nu = 0.17\ eV$

4.16 Let n be the number of photons per sec. per cm^2 incident on the photocathode.

$$n = \frac{10 \times 10^{-3}}{h\nu} = 7.77 \times 10^{16}\ \text{photon}/cm^2 \cdot sec$$

Current is:

$$I = \eta \cdot n \cdot e = 0.1 \times 7.77 \times 10^{16} \times 1.6 \times 10^{-19}$$

$$= 1.24 \times 10^{-3}\ A.$$

4.17 To make the cathode shot noise coming from the local oscillator equals to that of the dark current, Let:

$$\frac{P_L e\eta}{h\nu} = \bar{i}_d$$

$$P_L = \frac{\bar{i}_d \cdot h \cdot c}{e\eta\lambda} = 1.17 \times 10^{-7} \cdot \frac{1}{\eta\lambda}\ W.$$

If $\eta = 0.1$, $\lambda = 633\ nm$, then

$$P_L = 1.95\ W.$$

Chapter 5

5.1

$$W = 3.6 \text{ mm}, \quad \tau = 450 \text{ nsec}$$

$$V_A = \frac{W}{\tau} = 8000 \text{ m/sec}$$

5.2 (a).

$$f_{max} = \frac{1}{\tau} = \frac{1}{450 \text{ nsec}} = 2.22 \text{ MHz}.$$

(b). Focus the laser spot at the AO cell aperture can reduce the rise time.

To achieve 1 GHz bandwidth:

$$W = V_A \cdot \tau = 8000 \frac{\text{m}}{\text{sec}} \times 10^{-9} \text{ sec}.$$

$$= 8 \ \mu\text{m}.$$

5.3 When AO cell is driven by a chirp signal, the output beam scans.

5.4 set the axis of the analyzer perpenticular to that of the polarizer.

5.5 Set the axis of the analyzer perponticulat to that of the polarizer.

$$T/T_0 = \frac{\sin^2 20^\circ}{\sin^2 45^\circ} = 23.4\%$$

5.6 Yes.

Contrast will be reversed if analyzer is rotated by 90°.

5.7 Let $\Delta\phi = (2m+1)\pi$

$$\frac{(n_1 - n_2)\cdot 2d}{\lambda} = (m + \tfrac{1}{2})$$

$$E^2 = \frac{V_h^2}{d^2} = \frac{(m+\frac{1}{2})}{kd}$$

$$\therefore \quad V_h = \sqrt{(m+\tfrac{1}{2})d/2k}$$

5.8 No. Since V_h is not dependent on λ.

5.9 (a). No modulation

(b) polarization of the readout beam should be in 45° with respect of the PROM axis.

5.10

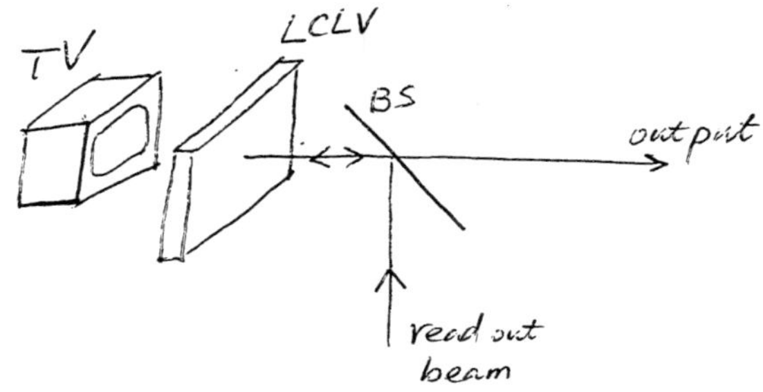

5.11 No solution is provided.

5.12 No solution is provided.

5.13

$$A = 10 \times 10 \ mm^2 = 1 \ cm^2 \ ; \quad \tau = 10 \ msec.$$

$$E = W \cdot \tau = 10 \ \mu J/cm^2$$

$$W = \frac{E}{\tau} = 1 \ mW.$$

5.14 If the response time of liquid crystal is comparable with frame speed (~ 30/sec.), Fourier transform can be obtained.

If the response time is much less than the frame speed, no Fourier transform can be observed.

5.15 (a) No.

(b). Yes. Phase modulation can be achieved without polarizer.

5.16 $I_{in} = \frac{E}{\tau} = \frac{25\ nJ/cm^2}{10\ msec} = 2.5\ \mu J/cm^2$

$I_{out} = I_{in} \cdot G = 250\ mJ/cm^2$

5.17 The birefregence is wavelength dependent. At given bias voltage and input intensity, the modulation of MSLM to different wavelength components is different. Therefore a color image is obtained.

5.18 Because it modulates only the phase.

5.19 $\eta_{max} = 33.9\%$

(Obtained by look up the Bessel function table)

5.20 Using spatial filtering technique, as discussed in example 5.5

5.21

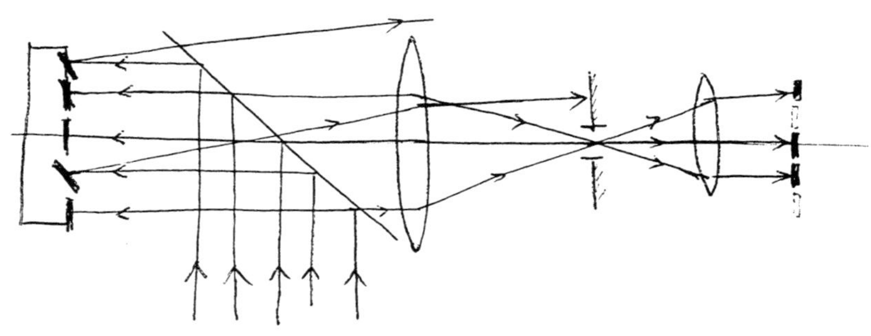

5.22 No solution is provided. Discussion among students is encouraged.

5.23 If the along-track pixel spacing is also 1.6μm,

$$N = \frac{1\ mm \times 1\ mm}{1.6\mu m \times 1.6\mu m} = 625 \times 625 \text{ pixels.}$$

5.24

$$2d\eta = \lambda$$

$$d = \frac{\lambda}{2\eta} = \frac{0.78}{2 \times 1.5} = 0.26\ \mu m$$

5.25 (a). $\Delta\eta_{max} = 4 \times 10^{-3}$, $\Delta\eta = 10^{-6}$

From Eq. (5.5)

$$t = -\tau \ln\left(1 - \frac{\Delta\eta}{\Delta\eta_{max}}\right)$$

$$= 1.25 \text{ msec.}$$

(b). $$E = \frac{2\Delta\eta}{\eta_o^3\, r_{eff}}$$

(c). $$E_{max} = \frac{2\Delta\eta_{max}}{\eta_o^3\, r_{eff}}$$

5.26 (a) $d \cdot n_0 = \frac{\lambda}{2}$

$$d = \frac{\lambda}{2n_0} = \frac{670}{2 \times 2.29} = 146.3 \text{ nm.}$$

(b) space shift $= \frac{d}{4} = 36.6$ nm

(c) average charge transport distance $= \frac{d}{2} = 73.1$ nm

(d) $$E_{max} = \frac{2\Delta n}{n_0} \cdot \frac{1}{n_0^2 r_{eff}} = \frac{2 \times 4 \times 10^{-3}}{2.29^2 \times 30.8 \times 10^{-12}}$$

$$= 4.95 \times 10^7 \text{ V/m}.$$

5.27 If the Bragg condition is satisfied, the diffraction efficiency can approach 100%.

Chapter 6

6.1 From Eq. (6.4).

$$\nu \geq \frac{W}{h} = \frac{1.6\times10^{-19}}{6.63\times10^{-34}} = 2.41\times10^{16}\ Hz.$$

6.2 From the equation given by example 8.1.

$$\lambda = \frac{1.24}{E} = 0.62\ \mu m.$$

It can be detected by a human eye.

6.3

$\lambda = 488$ nm $\longrightarrow$ $E = 2.54$ eV

$\lambda = 514$ nm $\longrightarrow$ $E = 2.41$ eV

$\lambda = 632.8$ nm $\longrightarrow$ $E = 1.96$ eV

$\lambda = 1.06$ μm $\longrightarrow$ $E = 1.17$ eV

6.4 From example 6.2

$$\frac{N_2}{N_1} = \frac{B_{12}(\nu)\,\rho(\nu)}{B_{12}(\nu)\cdot\rho(\nu) + \frac{1}{t_{sp}}} \approx B_{12}\,\rho(\nu)\,t_{sp}$$

$$= \frac{1}{\exp(h\nu/kT) - 1} \approx \exp(-h\nu/kT)$$

$$= 1.64\times10^{-17}$$

6.5 From Eqs. (6.6) and (6.7)

$$I_{sp} = I(0) \cdot e^{-\frac{t}{t_{sp}}}$$

$$I_{st} = I(0) \cdot e^{-\beta_{12} \rho(\nu) t}$$

$$\frac{I_{sp}}{I_{st}} = e^{-t \left[\frac{1}{t_{sp}} + \beta_{12} \rho(\nu) \right]} = e^{-t_{sp} \cdot t \left[1 + \beta_{12} \rho(\nu) t_{sp} \right]}$$

$\because \quad \beta_{12} \rho(\nu) t_{sp} \ll 1$

$$\therefore \quad \frac{I_{sp}}{I_{st}} \approx e^{-t_{sp} \cdot t}$$

6.6

$$\beta_{12} \rho(\nu) = \frac{1}{t_{sp}} \cdot \frac{1}{e^{\frac{h\nu}{KT}} - 1}$$

$$= 10^{6} \cdot \frac{1}{e^{\frac{(1.24/0.6) \times 1.6 \times 10^{-19}}{300 \times 1.38 \times 10^{-23}}} - 1}$$

$$= 2.053 \times 10^{-29}$$

6.7

$$\lambda_{pump} = \frac{1.24}{E_3 - E_1} = \frac{1.24}{3.1} = 0.4 \ \mu m$$

Xenon lamp can be used.

6.8

$$E_{pump} = \frac{1.24}{\lambda_{pump}} = \frac{1.24}{0.4} = 3.1 \ eV$$

$$E_{laser} = \frac{1.24}{\lambda_{laser}} = \frac{1.24}{0.694} = 1.784 \text{ eV}$$

$$\eta = \frac{E_{laser}}{E_{pump}} = \frac{1.784}{3.1} = 57.5\%$$

6.9 No solution available.

6.10 No solution available. Discussion among students is encouraged.

6.11 $x = c \cdot t = 3 \times 10^{10} \times 10^{-6} = 3 \times 10^{4}$ cm

$$\frac{I(x)}{I_0} = e^{Gx} = e^{0.005 \times 3 \times 10^{4}} = 1.39 \times 10^{65}$$

6.12 From (6.13)

$$L \geq -\frac{1}{2G} \ln(R_1 R_2)$$

$$= -\frac{1}{2 \times 0.005} \ln(0.99)$$

$$= 1 \text{ cm.}$$

6.13 Let:

$$100 \times 300 \times 2\theta < 2/2$$

$$\theta < \frac{1}{60000} = 1.67 \times 10^{-5} \text{ rad}$$

$$= 3.44''$$

6.14 $$\Delta\lambda = \frac{\lambda^2}{2\eta L}$$

$$L_{max} = \frac{\lambda^2}{2\eta \cdot \Delta\lambda_G} = \frac{0.6328^2}{2 \times 2 \times 10^{-3}}$$

$$= 100 \text{ mm.}$$

6.15 $$\frac{dw}{dz} = \frac{1}{2w} \cdot \frac{2w_0 z}{z_0^2}$$

$$= \frac{w_0^3 z}{z_0^2} \left(1 + \frac{z^2}{z_0^2}\right)$$

6.16 No solution available. Discussion among students is encouraged.

6.17 No solution available. Discussions are encouraged.

6.18 $$l_c = \frac{\lambda^2}{2\Delta\lambda} = 0.15 \text{ mm}$$

6.19 No solution available. Discussion is encouraged.

6.20 $$\sin\theta \approx \frac{\lambda}{d} = \frac{0.85}{2} = 0.425$$

$$\theta \approx 25^\circ$$

6.21 No solution available.

Chapter 7

7.1

We know that the Fourier transform of an exponential attenuation function

$[\ a(x)=0,\ \text{when } x<0;\ =e^{-x},\ \text{when } x\geq 0\ ]$ is

$$A(p)=\int_{-\infty}^{+\infty} a(x)\,e^{-ipx}\,dx=\int_{0}^{+\infty} e^{-x}\,e^{-ipx}\,dx$$

$$=\int_{0}^{+\infty} e^{-x(1+ip)}\,dx=\frac{1}{1+ip}.$$

Thus the spatial impulse response of the given system is the inverse Fourier transform of the transfer function:

$$h(x)=\mathcal{F}^{-1}[H(p)]=\mathcal{F}^{-1}\left[\frac{ip}{1+ip}\right]=\mathcal{F}^{-1}\left[\frac{1+ip-1}{1+ip}\right]$$

$$=\mathcal{F}^{-1}\left[1-\frac{1}{1+ip}\right]=\mathcal{F}^{-1}(1)-\mathcal{F}^{-1}\left(\frac{1}{1+ip}\right)=\delta(x)-a(x)$$

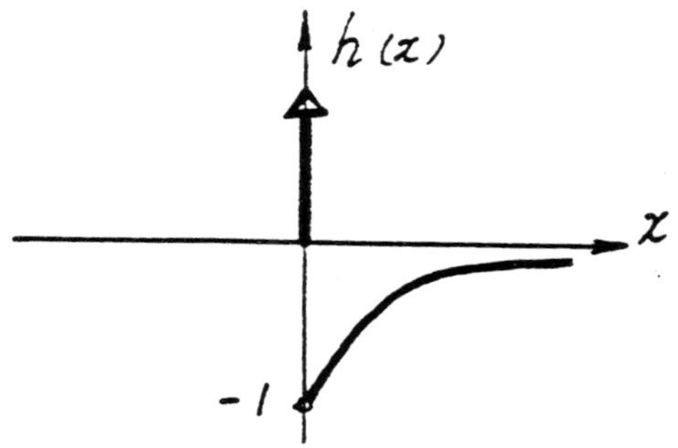

7.2 In the calculation of $h(x)$ in Problem 7.1, we

saw that

(a) $\mathcal{F}^{-1}\left[1-\frac{1}{1+ip}\right] = \mathcal{F}^{-1}[1] + \mathcal{F}^{-1}\left[-\frac{1}{1+ip}\right]$,

which means $\mathcal{F}^{-1}[F_1(p)] + \mathcal{F}^{-1}[F_2(p)] = \mathcal{F}^{-1}[F_1(p) + F_2(p)]$,

(b) $\mathcal{F}^{-1}\left[-\frac{1}{1+ip}\right] = -\mathcal{F}^{-1}\left[\frac{1}{1+ip}\right]$

which means $\mathcal{F}^{-1}[K F_3(p)] = K\,\mathcal{F}^{-1}[F_3(p)]$.

so both additivity and homogeneity properties hold in problem 7.1.

7.3

In a linear system, which impulse response is $h(x)$, the output excitation $g(x)$ is

$$g(x) = f(x) * h(x), \quad \text{where } f(x) \text{ is the input.}$$

Now the input is translated, the new output excitation is

$$g'(x) = f(x-x_0) * h(x) = \int f(\alpha - x_0)\, h(x-\alpha)\, d\alpha$$

$$= \int f(\alpha - x_0)\, h(x - x_0 - \alpha + x_0)\, d\alpha$$

$$= \int f[(\alpha - x_0)]\, h[(x-x_0) - (\alpha - x_0)]\, d(\alpha - x_0)$$

$$= \int f(y)\, h[(x-x_0) - y]\, dy = g(x - x_0)$$

7.4

The Fourier transform of the input is

$$F(p) = \mathcal{F}[f(x)] = \frac{1}{2}\left[\delta(p-p_1) + \delta(p+p_1) + \delta(p-p_2) + \delta(p+p_2)\right]$$

the Fourier transform of the output excitation is

$$G(p) = F(p)\, H(p)$$

due to the property of δ-function:

$$f(x)\,\delta(x-x_0) = f(x_0)\,\delta(x-x_0)$$

$$G(p) = \frac{1}{2}\left[e^{-ip_1^2}\delta(p-p_1) + e^{-i(-p_1)^2}\delta(p+p_1) + e^{-ip_2^2}\delta(p-p_2) + e^{-i(-p_2)^2}\delta(p+p_2)\right]$$

$$= \frac{1}{2}\, e^{-ip_1^2}\left[\delta(p-p_1) + \delta(p+p_1)\right] + \frac{1}{2}\, e^{-ip_2^2}\left[\delta(p-p_2) + \delta(p+p_2)\right]$$

So the output excitation

$$g(x) \quad \mathcal{F}^{-1}[G(p)] \quad e^{-ip_1^2}\cos p_1 x \quad e^{-ip_2^2}\cos p_2 x$$

From this result, we see that different coefficients are given to different components of the input.

7.5 The given transfer function may be expressed as

$$H(p) = \mathrm{rect}\left(\frac{p}{p_c}\right) \cdot e^{-i\alpha p}$$

the Fourier transform of the output response is

$$G(p) = F(p) \cdot H(p) = \mathcal{F}[\delta(x) - \delta(x-x_0)] \cdot H(p)$$

$$= (1 + e^{-ix_0 p}) \cdot \mathrm{rect}\left(\frac{p}{p_c}\right) e^{-i\alpha p}$$

the output is its inverse Fourier transform,

$$g(x) = \mathcal{F}^{-1}[G(p)] = \mathrm{sinc}[p_c(x-\alpha)] + \mathrm{sinc}[p_c(x-x_0-\alpha)]$$

7.6

The input is expressed by $f(x) = \mathrm{rect}\left(\frac{x}{\Delta x}\right)$, its Fourier transform is $F(p) = \Delta x \ \mathrm{sinc}(\Delta x\, p)$. the Fourier transform of the output response is

$$G(p) = H(p) \cdot F(p) = \mathrm{rect}\left(\frac{p}{p_c}\right) e^{-i\alpha p} \cdot \Delta x \ \mathrm{sinc}(\Delta x \cdot p)$$

a. when $\Delta x \ll \frac{\pi}{p_c}$, $G = H \cdot F \approx$, $g = \mathcal{F}^{-1}[G] = \mathcal{F}^{-1}[\ ] =$ (x)

SPATIAL DOMAIN $f(x)$ x * $h(x)$ α x = $g \approx h$ α x

F.T. ⇓ FT ⇓ FT ⇑ FT^{-1}

FREQUENCY DOMAIN $F(p)$ p · $H(p)$ p = $G(p)$ $p_c/2$ p

From a point object, we get a impulse response, shifted

b. when $\Delta x \gg \frac{\pi}{p_c}$, $G = H \cdot F \approx H$, $g(x) = \mathcal{F}^{-1}[G] = \mathcal{F}^{-1}[H] = h(x)$

S.D. $f(x)$ x * $h(x)$ α x = $g = f(x-\alpha)$ α

F.T. ⇓ F.T. ⇓ F.T. ⇑ FT^{-1}

F.D. $F(p)$ p · (p) p = $G(p)$ $p_c/2$

The object is quite broad, the spread relatively small, only shift.

c. when $\Delta x = \frac{\pi}{p_c}$, due to the high frequency is cut off, also phase modulated, the output is both shifted and smoothed.

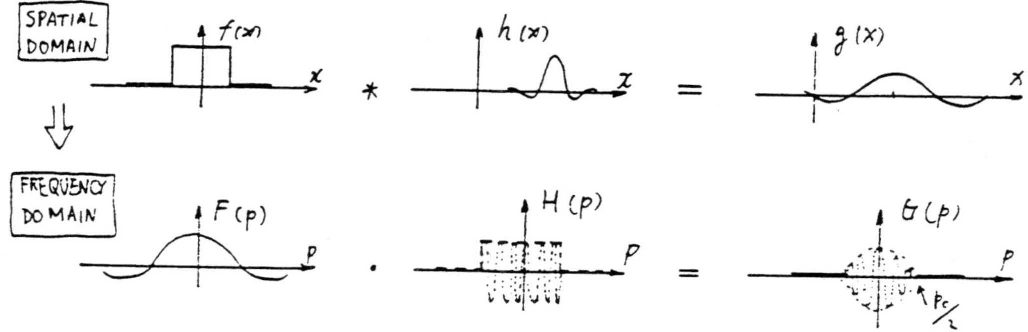

7.7

a. $f(x) = \sin p_1 x + \cos p_2 x = \frac{1}{2j}(e^{ip_1x} - e^{-ip_1x}) + (e^{ip_2x} + e^{-ip_2x})$

Its Fourier transform is

$$F(p) = \frac{1}{2j}[\delta(p-p_1) - \delta(p+p_1)] + \frac{1}{2}[\delta(p-p_2) + \delta(p+p_2)]$$

please refer the solution to 7.7 (b)

b. $f(x) = e^{ip_0x} + e^{-ip_0x} = 2\cos p_0 x$

$$F(p) = \int (e^{ip_0x} + e^{-ip_0x})\, e^{-ipx}\, dx$$

$$= \int e^{-ix(p-p_0)} dx + \int e^{-ix(p+p_0)} dx$$

$$= \delta(p-p_0) + \delta(p+p_0)$$

c. $f(x) = \sum_{n=-\infty}^{\infty} \delta(x - n\alpha)$

$$F(p) = \int [\Sigma\, \delta(x-n\alpha)]\, e^{-ipx}\, dx$$

$$= \Sigma \int \delta(x-n\alpha)\, e^{-ipx}\, dx = \Sigma\, e^{-in\alpha p}$$

$$= \sum_{m=-\infty}^{\infty} \delta(p - \frac{n}{\alpha})$$

7.8 $f_1(x)=\frac{1}{2}(e^{ip_1x}+e^{-ip_1x})$, its Fourier spectrum is

$$F_1(p)=\int_{-\infty}^{+\infty} f_1(x)e^{-ipx}dx=\int_{-\infty}^{+\infty}\frac{1}{2}(e^{ip_1x}+e^{-ip_1x})e^{-ipx}dx$$

$$=\int_{-\infty}^{+\infty}\frac{1}{2}e^{-i(p-p_1)x}dx+\int_{-\infty}^{+\infty}\frac{1}{2}e^{-i(p+p_1)x}dx$$

$$=\frac{1}{2}\delta(p-p_1)+\frac{1}{2}\delta(p+p_1)$$

$f_2(x)=\frac{1}{2i}(e^{ip_1x}-e^{-ip_1x})$, its Fourier spectrum is

$$F_2(p)=\int_{-\infty}^{+\infty} f_2(x)e^{-ipx}dx=\int_{-\infty}^{+\infty}\frac{1}{2i}(e^{ip_1x}-e^{-ip_1x})e^{-ipx}dx$$

$$=\int_{-\infty}^{+\infty}\frac{1}{2i}e^{-i(p-p_1)x}dx-\int_{-\infty}^{+\infty}\frac{1}{2i}e^{-i(p+p_1)x}dx$$

$$=\frac{1}{2i}\delta(p-p_1)-\frac{1}{2i}\delta(p+p_1)$$

Let $g_1(x)=e^{ip_1x}$, its Fourier transform is $G_1(p)=\delta(p-p_1)$

$g_2(x)=e^{-ip_1x}$, its Fourier transform is $G_2(p)=\delta(p+p_2)$

from the results we obtained, i.e.

$f_1(x)=\frac{1}{2}g_1(x)+\frac{1}{2}g_2(x)$, $F_1(p)=\frac{1}{2}G_1(p)+\frac{1}{2}G_2(p)$

and $f_2(x)=\frac{1}{2i}g_1(x)-\frac{1}{2i}g_2(x)$, $F_2(p)=\frac{1}{2i}G_1(p)-\frac{1}{2i}G_2(p)$

we may conclude that

if $g_1\to G_1$, $g_2\to G_2$, then $c_1g_1+c_2g_2\to c_1G_1+c_2G_2$

so Fourier transformation is a linear one.

7.9 Its Fourier transform is

$$F(p) = \int_{-\infty}^{+\infty} f(x)\, e^{-ipx}\, dx = \int_{0}^{+\infty} e^{-\alpha x}\, e^{-ipx}\, dx$$

$$= \int_{0}^{+\infty} e^{-x(\alpha+ip)}\, dx = \frac{1}{\alpha+ip}\, e^{-x(\alpha+ip)}\Big|_{x=\infty}^{x=0}$$

$$= \frac{1}{\alpha+ip}$$

7.10 If $f(x) \xrightarrow{F.T.} F(p)$, $f(x)$ is shifted,

then $\left|\mathcal{F}[f(x-x_0)]\right| = \left|\int_{-\infty}^{+\infty} f(x-x_0)\, e^{-ipx}\, dx\right|$

$$= \left|\int_{-\infty}^{+\infty} f(x-x_0)\, e^{-ip(x-x_0+x_0)}\, d(x-x_0)\right|$$

$$= \left|\int_{-\infty}^{+\infty} f(x-x_0)\, e^{-ip(x-x_0)} \cdot e^{-ipx_0}\, d(x-x_0)\right|$$

$$= \left|\left[\int_{-\infty}^{+\infty} f(\xi)\, e^{-ip\xi}\, d\xi\right] e^{-ipx_0}\right| = \left|F(p) \cdot e^{-ipx_0}\right|$$

$$= |F(p)|\,|e^{-ipx_0}| = |F(p)| \cdot 1 = |\mathcal{F}[f(x)]|.$$

7.11 If $f(x) \xrightarrow{F.T.} F(p)$ (F.T. = Fourier Transform)

after amplitude modulated

$$g(x) = f(x)\cos p_0 x = \frac{f(x)}{2}\left[e^{ip_0x} + e^{-ip_0x}\right]$$

The Fourier spectrum of $g(x)$ is

$$G(p) = \int_{-\infty}^{+\infty} g(x)\, e^{-ipx}\, dx$$

$$= \int \frac{f(x)}{2} e^{ip_0 x} e^{-ipx} dx + \int \frac{f(x)}{2} e^{-ip_0 x} e^{-ipx} dx$$

$$= \frac{1}{2} \int f(x) e^{-ix(p-p_0)} dx + \frac{1}{2} \int f(x) e^{-i(p+p_0)x} dx$$

$$= \frac{1}{2} F(p-p_0) + \frac{1}{2} F(p+p_0)$$

This may be expressed by the following figue:

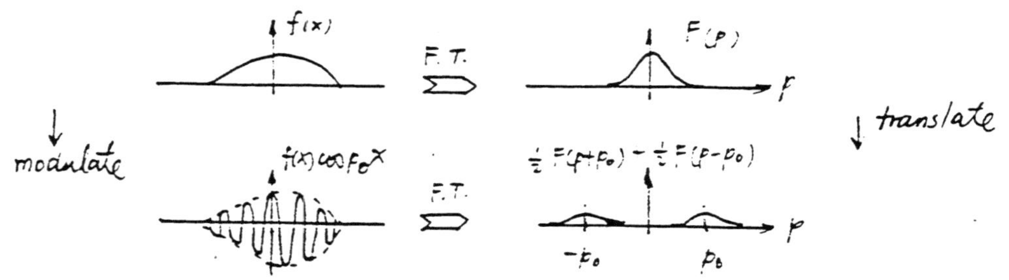

7.12 a) The sampled signal may be expressed as

$$g(x) = \sum_{n=-\infty}^{+\infty} f_n (x-nd) = \sum f(x)\,\delta(x-nd)$$

$$= f(x) \cdot \Sigma\, \delta(x-nd) \quad \text{or} \quad f(x) \cdot \text{comb}\,\frac{x}{d}.$$

Refer to §7.5. Property of Fourier Transformation, 4. convolution property, the Fourier Transform of $g(x)$ is

$$G(p) = F(p) * \mathcal{F}[\text{comb}(\frac{x}{d})]$$

$$= F(p) * \Sigma\, \delta(p - \frac{n}{d})$$

$$= \Sigma\, F(p - \frac{n}{d})$$

This process may also be depicted as following figue:

IN SPATIAL DOMAIN
SAMPLING

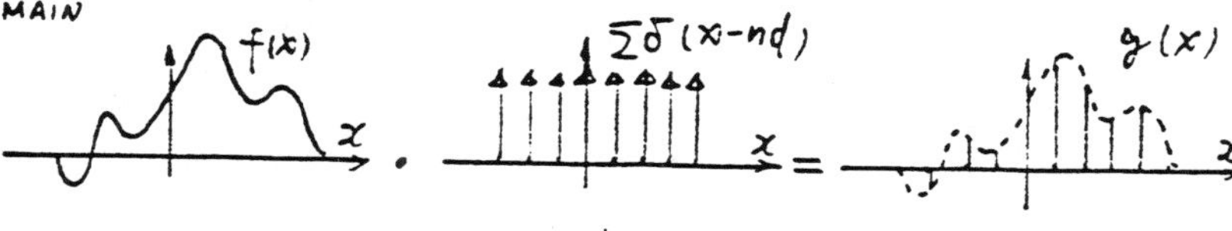

⇓ F.T. ⇓ F.T.

IN FREQUENCY DOMAIN
REPEATING

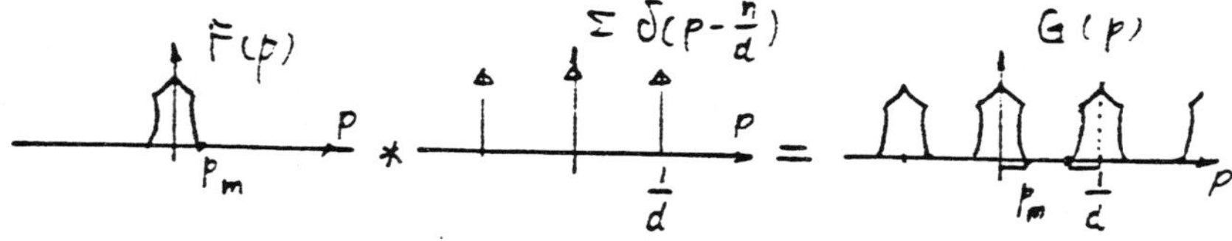

b) To avoid the overlapping of adjacent peaks, $\frac{1}{d}$ should be more than $2p_m$

$$\frac{1}{d} > 2p_m \quad , \quad \text{or} \quad d < \frac{1}{2p_m}$$

This is shown in the figue.

7.13

The output is the convolution of $f(x)$ and $h(x)$: $g(x) = f(x) * \quad (x)$

According to the convolution property of Fourier transformation,

$$G(p) = F(p) \cdot H(p)$$

According to scale property of Fourier transform (see §7.5.3.

$f(x) = \text{rect}\left(\frac{x}{\Delta x}\right)$, $\Rightarrow F(p) = \Delta x \ \text{sinc}\ \Delta x \cdot p$

$h(x) = \frac{p_c}{\pi} \frac{\sin(p_c x)}{(p_c x)} \Rightarrow H(p) = \text{rect}\left(\frac{p}{p_c}\right)$

If $\Delta x >> \frac{2\pi}{p_c}$, or $p_c >> \frac{2\pi}{\Delta x}$, most frequency components of $F(p)$ can pass the given system,

$G(p) \approx F(p)$

so $g(x) \approx f(x)$. Please refer the following figure:

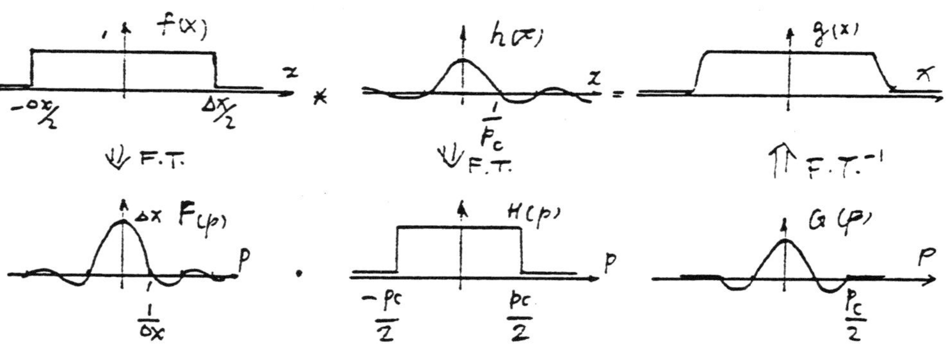

7.14 If $f(x) \xrightarrow{\text{F.T.}} F(p)$

the the inverse Fourier transform of $F^*(p)$ is $\frac{1}{2\pi}\int F^*(p)\, e^{ipx}\, dp = \left[\frac{1}{2\pi}\int F(p)\, e^{-ipx}\, dp\right]^*$

$$= \left[\frac{1}{2\pi}\int F(p)\, e^{ip(-x)}\, dp\right]^* = [\, f(-x)]^* = f^*(-x)$$

7.15

a) The impulse response of the given system is $h(x) = \mathcal{F}^{-1}[H(p)] = \mathcal{F}^{-1}[F^*(p)]$

according to the answer of Problem 1.14

$$h(x) = f^*(-x)$$

so the output is

$$g(x) = f(x) * h(x) = f(x) * f^*(-x) = R_{ff}(x).$$

$R(x)$ is the autocorrelation.

b) When the input signal is $u(x)$, the output is

$$k(x) = u(x) * h(x) = u(x) * f^*(-x) = R_{fu}(x).$$

This is the cross correlation.

c) From the viewpoint of filtering, in frequency domain, the spectrums of input signals are

$f(x) \to F(p)$, $u(x) \to U(p)$, respectively,

after filtering, we get

$F(p)\,H(p) = F(p)\,F^*(p) = |F(p)|^2$ for (a).

$U(p)\,H(p) = U(p)\,F^*(p)$ for (b).

so the output results are

$|F(p)|^2 \xrightarrow{FT^{-1}} R_{ff}(x)$, for (a);

$U(p)F^*(p) \xrightarrow{FT^{-1}} R_{fu}(x)$, for (b);

respectively.

The significance of the Wiener-Khintchin theorem is that it predicts the output from the matched filter.

7.16

The pulse train may be expressed as

$$g(x) = \Sigma \, \text{rect}\, \frac{x-nT}{\Delta x} = \text{rect}\, \frac{x}{\Delta x} * \Sigma\, \delta(x-nT)$$

According to the convolution property of Fourier transformation, its frequency spectrum is

$$G(p) = \mathcal{F}[\text{rect}(\tfrac{x}{\Delta x})] \cdot \mathcal{F}[\Sigma\, \delta(x-nT)]$$
$$= \Delta x \,\text{sinc}(\Delta x\, p) \cdot \Sigma\, \delta(p - \tfrac{n}{T}).$$

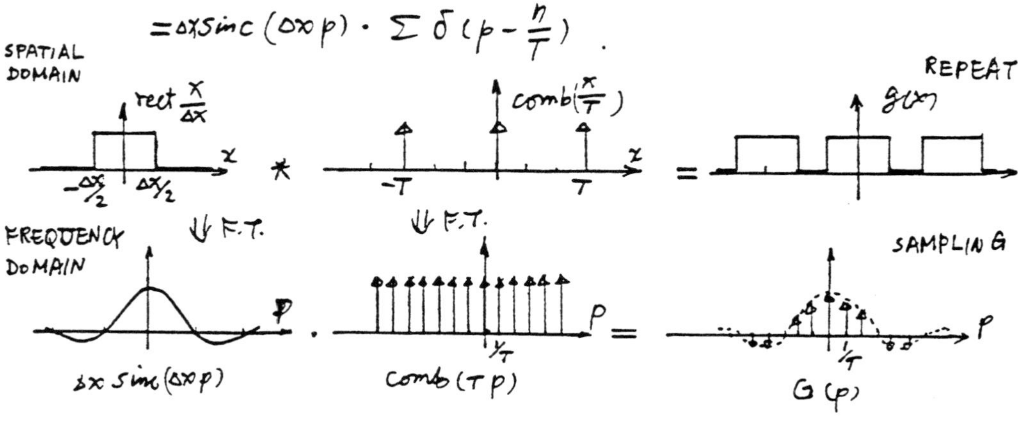

Chapter 8

8.1 Refer to Fig 8.2 and Eq (8.2), let $\Delta l = \frac{\lambda}{20}$, the distance between the diffraction screen and the observation plane

$$l = \frac{\rho^2}{2\Delta l} = \frac{(5/2)^2}{2\times 6.3\times 10^{-5}\div 20} = 9.92\times 10^4 \text{ mm} \text{ or } 99.2 \text{ m}.$$

8.2 The transmittance of the circular aperture may be expressed as

$$T(x,y) = \mathrm{circ}\left(\frac{r}{r_0}\right) = \begin{cases} 1 & r \leq r_0 \\ 0 & r > r_0, \end{cases} \text{ where } r_0 = \frac{D}{2}$$

The light field immediately behind the diffraction screen is

$$f(x,y) = A\ \mathrm{circ}\left(\frac{r}{r_0}\right),$$

A is the amplitude of the illuminating monochromatic plane wave.

According to Eq (8.19), the diffraction light field at observation plane is

$$g(\alpha,\beta) = C\iint f(x,y)\, e^{[-\frac{ik}{l}(\alpha x+\beta y)]}\, dx\, dy$$

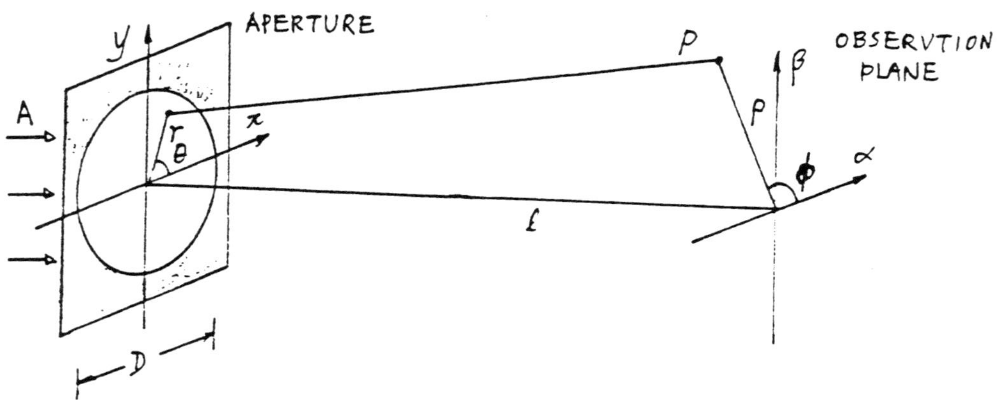

Sutstitute the coordinates transform formula

$$\begin{cases} x = r\cos\theta \\ y = r\sin\theta \end{cases} \qquad \begin{cases} \alpha = \rho\cos\phi \\ \beta = \rho\sin\phi \end{cases}$$

into $g(\alpha,\beta)$:

$$g(\alpha,\beta) = C\int_0^{\infty}\int_0^{2\pi} f(r,\theta)\, e^{-j\frac{2\pi}{\lambda\ell} r\rho\cos(\theta-\phi)}\, r dr d\theta$$

Choose a point P on α- axis, $\phi = 0$

$$g(\alpha,\beta) = C\int_0^{r_0}\int_0^{2\pi} e^{-j\frac{2\pi}{\lambda\ell} r\rho\cos\theta}\, r dr\, d\theta$$

To calculate this integral, let $Z = \frac{2\pi}{\lambda\ell}\rho r$,

then $r = \frac{Z}{\frac{2\pi}{\lambda\ell}\rho}$, $dr = \frac{dz}{\frac{2\pi}{\lambda\ell}\rho}$,

$$g(z) = \int_0^{Z_0} \left(\frac{\lambda\ell}{2\pi\rho}\right)^2 \left[\int_0^{2\pi} e^{-jZ\cos\theta}\, d\theta\right] Z\, dZ$$

$$= \int_0^{Z_0} \left(\frac{\lambda\ell}{2\pi\rho}\right)^2 2\pi J_0(Z)\, Z\, dZ$$

$$= \frac{1}{2\pi}\left(\frac{\lambda\ell}{\rho}\right)^2 \int_0^{Z_0} Z\, J_0(Z)\, dZ$$

$$= 2\pi \left(\frac{\lambda l}{2\pi\rho}\right)^2 Z_0 J_1(Z_0)$$

or expressed as

$$g(\rho) = 2\pi \left(\frac{\lambda l}{2\pi\rho}\right)^2 \left(\frac{2\pi\rho}{\lambda} r_0\right) J_1\left(\frac{2\pi\rho}{\lambda} r_0\right) = 2\pi r_0^2 \frac{J_1(Z_0)}{Z_0}$$

$g(\rho)$ is a circular symmetrical function, zero intensity happens at $Z_0 = 3.8317, 7.0156, \cdots$ So the radius of the first dark ring is ρ_1

$$\frac{2\pi r_0}{\lambda l}\rho_1 = 3.8317, \quad \rho_1 = \frac{3.8317}{\pi} \cdot \frac{\lambda l}{2 r_0} = \frac{1.22\lambda}{\frac{D}{l}}$$

The intensity distribution of the Fraunhofer diffraction pattern from a circular aperture is also known as Airy disc.

8.3 The light field immediately behind the diffraction screen is

$$f_{array}(x_1, y_1) = f(x_1, y_1) + f(x_1 - d, y_1)$$

the diffraction field due to $f(x_1, y_1)$ is

$$g(x, y) = C \iint f(x_1, y_1)\, e^{-\frac{ik}{l}(x x_1 + y y_1)} dx_1 dy_1$$

the diffraction field due to $f(x_1-d, y_1)$ is

$$g_d(x, y) = C \iint f(x_1-d, y_1)\, e^{-i\frac{k}{l}(xx_1+yy_1)}\, dx_1\, dy_1$$

$$= C \iint f(x_1-d, y_1)\, e^{-i\frac{k}{l}[x(x_1-d+d)+yy_1]}\, dx_1\, dy_1$$

$$= C \iint f(x_1-d, y_1)\, e^{-i\frac{k}{l}[x(x_1-d)+yy_1]}\, e^{-i\frac{k}{l}dx}\, dx_1\, dy_1$$

$$= C\, e^{-i\frac{kxd}{l}}\, g(x, y)$$

Thus, the light intensity distribution on observation screen is

$$I(x,y) = |g(x,y) + g_d(x,y)|^2 = |g(x,y)|^2 + |g_d(x,y)|^2 +$$

$$+ g^* g_d + g g_d^* = |g(x,y)|^2 + |g(x,y)|^2 +$$

$$+ |g(x,y)|^2 e^{-i\frac{kxd}{l}} + |g(x,y)|^2 e^{i\frac{kxd}{l}}$$

$$= |g(x,y)|^2 \left[2 + 2\cos\frac{kxd}{l}\right] = 4|g(x,y)|^2 \cos^2\frac{kxd}{2l}$$

Spacing $\Delta = \frac{\lambda l}{d}$

8.4

a)

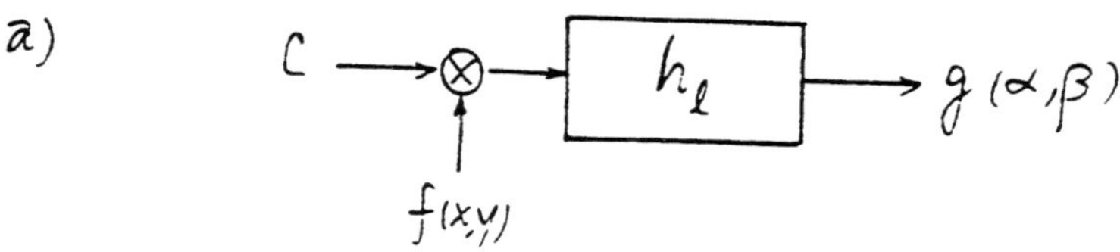

b) $g(\alpha,\beta) = C\, f(x,y) * h_\ell$

$$= C \iint f(x,y)\, e^{i\frac{k[(x-\alpha)^2+(y-\beta)^2]}{2\ell}}\, dx\, dy$$

8.5

a)

$u(x,y) \longrightarrow \otimes \longrightarrow$ [h_ℓ] $\longrightarrow g(\alpha,\beta)$, with $f(x,y)$ entering the multiplier $\otimes$

b) $g(\alpha,\beta) = [u(x,y)\, f(x,y)] * h_\ell$

$$= \iint u(x,y)\, f(x,y)\, e^{i\frac{k[(x-\alpha)^2+(y-\beta)^2]}{2\ell}}\, dx\, dy$$

8.6

a)

$f(x,y) \longleftarrow$ [h_ℓ^*] $\longleftarrow g(\alpha,\beta)$

b) $g(\alpha,\beta) * h_\ell^* = [f(x,y) * h_\ell] * h_\ell^*$

$$= f(x,y) * [h_\ell * h_\ell^*] = f(x,y) * \delta(x,y) = f(x,y)$$

APPENDIX :

Please notice

$$h_\ell * h_\ell^* = \iint e^{i\frac{k(\alpha^2+\beta^2)}{2\ell}} \cdot e^{-j\frac{k[(\alpha-x)^2+(\beta-y)^2]}{2\ell}} d\alpha d\beta$$

$$= \iint [e^{i\frac{k(\alpha^2+\beta^2)}{2}} e^{-j\frac{k(\alpha^2+\beta^2)}{2}}] e^{-jk\frac{x^2+y^2}{2\ell}} e^{j\frac{k(\alpha x+\beta y)}{\ell}} d\alpha d\beta$$

$$= C \iint e^{j2\pi(\frac{\alpha x+\beta y}{\lambda\ell})} d\alpha d\beta = C\ \delta(x, y)$$

8.7

a)

$f(\alpha_1,\beta_1) \rightarrow$ [h_f] $\rightarrow u_1 \otimes u_2 \rightarrow$ [h_ℓ] $\rightarrow u_3 \otimes u_4 \rightarrow$ [h_f] $\rightarrow g(\alpha_2,\beta_2)$

(first multiplier input: $T_\ell(x_1, y_1)$; second multiplier input: $T_\ell(x_2, y_2)$)

b) $f(\alpha_1, \beta_1) = \delta(\alpha_1, \beta_1)$

$u_1 = f(\alpha_1, \beta_1) * h_f = \delta(\alpha_1, \beta_1) * e^{j\frac{k(x_1^2+y_1^2)}{2f}} = e^{j\frac{k(x_1^2+y_1^2)}{2f}}$ (divergent spherical)

$u_2 = u_1 \cdot T_\ell = e^{j\frac{k(x_1^2+y_1^2)}{2f}} \cdot e^{-j\frac{k(x_1^2+y_1^2)}{2f}} = 1$ (plane wave)

$u_3 = u_2 * h_\ell = C$ (plane wave)

$u_4 = u_3 \cdot T_\ell = C\, e^{-j\frac{k(x_2^2+y_2^2)}{2f}}$ (convergent spherical wave)

$u_5 = g(\alpha_2 \beta_2) = e^{-j\frac{k(x_2^2+y_2^2)}{2f}} * e^{j\frac{k(\alpha_2^2+\beta_2^2)}{2f}} = C\ \delta(\alpha_2, \beta_2)$

(please refer to the APPENDIX to Problem 8.6)

8.8

a) According to Huygens's principle the difference of the distances from adjacent slits to P must be λ, or $n \cdot \lambda$

$l_i = \sqrt{l_0^2 + h_i^2}$ →

By paraxial approximation, →

$l_i \approx l_0 \left(1 + \frac{h_i^2}{2l_0^2}\right)$

$h_i = \sqrt{2l_0(l_i - l_0)}$. Since the optical disturbances from those slits are in phase, $l_i - l_0 = n\lambda$, n is an arbitrary integer. Say $n = 1, 2, 3, 4, \cdots$ (or $2, 4, 5, 8, \cdots$) So there are many possibilities, the simplest one is $n = 1, 2, 3, 4$,

$h_1 = \sqrt{2 \times 1000\ \text{mm} \times 1 \times 0.6 \times 10^{-3}\ \text{mm}} = \sqrt{1.2}\ \text{mm} = 1.095\ \text{mm}$

$h_2 = \sqrt{2l_0 \cdot 2\lambda} = \sqrt{2} \cdot h_1 = 1.549\ \text{mm}$

$h_3 = \sqrt{2l_0 \cdot 3\lambda} = \sqrt{3} \cdot h_1 = 1.897\ \text{mm}$

$h_4 = \sqrt{2l_0 \cdot 4\lambda} = 2\ h_1 = 2.191\ \text{mm}$

the distances between neighboring slits are

$d_{1-0} = h_1 = 1.10\ \text{mm}$ $\quad d_{2-1} = h_2 - h_1 = 0.45\ \text{mm}$

$d_{3-2} = h_3 - h_2 = 0.35\ \text{mm}$ $\quad d_{4-3} = h_4 - h_3 = 0.29\ \text{mm}$

b) Under incoherent illumination,

light disturbance from each slit is u_o

then, the intensity at P is

$$u_o^2 + u_o^2 + u_o^2 + u_o^2 + u_o^2 = 5\, u_o^2 = I_{inc}$$

Under coherent illumination,

the intentensity at P becomes

$$(u_o + u_o + u_o + u_o + u_o)^2 = 25\, u_o^2 = I_c = 5\, I_{inc}.$$

(We neglect the intensity change due to l_i.)

8.9

(a) The transmittance of the given zoneplate may be expressed as

$$t(r) = \frac{1}{2} + \frac{1}{2}\ \text{sgn} \cos \alpha r^2$$

where $r = \sqrt{x^2 + y^2}$,

$$\text{sgn}(\alpha) = \begin{cases} 1, & \alpha > 0 \\ 0, & \alpha = 0 \\ -1, & \alpha < 0 \end{cases}$$

(b) let $z = r^2$

$$t(z) = \frac{1}{2} + \frac{1}{2}\ \text{sgn} \cos \alpha z$$

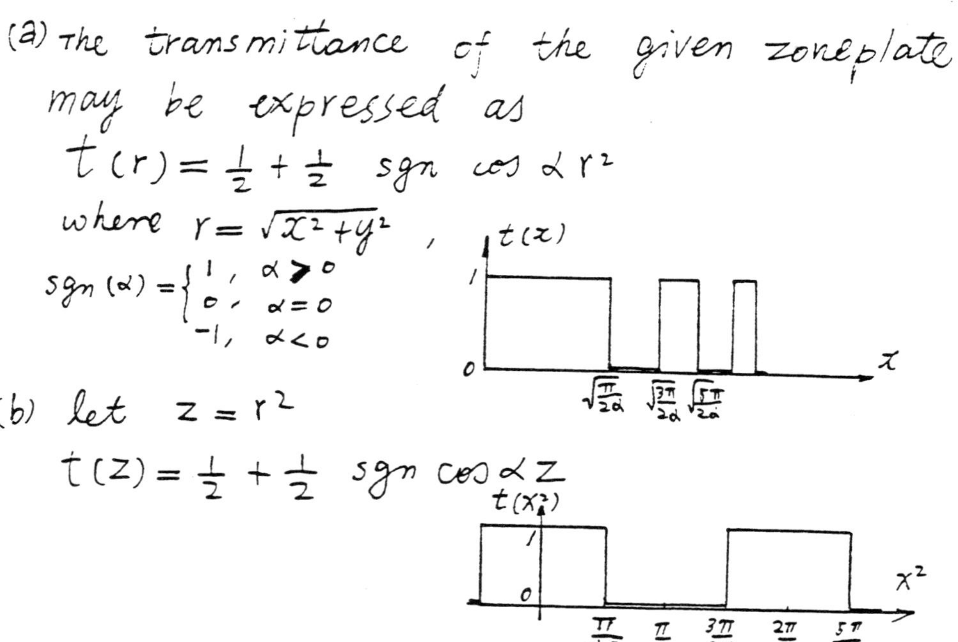

(C) The transmittance function may also be expressed as

$$t = \text{rect}\,\frac{2z}{d} * \frac{1}{d}\,\text{comb}\,\frac{z}{d}\ , \text{ where } d = \frac{2\pi}{\alpha}$$

Its Fourier transform is

$$T(p) = \frac{d}{2}\,\text{sinc}\,\frac{d}{2}p \cdot \Sigma\,\delta\left(p - \frac{n}{d}\right)$$

See solution to Problem 7.16, let $\Delta x = \frac{d}{2}$, $T = d$.

So $$t = \iint T(p)\, e^{jpx}\, dp$$

$$= \iint \frac{d}{2}\,\text{sinc}\left(\frac{d}{2}p\right) \Sigma\,\delta\left(p - \frac{n}{d}\right) e^{jpz}\, dp$$

$$= \sum_n \text{sinc}\left(\frac{d}{2} \cdot \frac{n}{d}\right) e^{j\frac{n}{d}z}$$

$$= \sum_n \text{sinc}\,\frac{n}{2}\; e^{jn\alpha r^2}$$

or $$\sum_{n=-\infty}^{+\infty} \frac{\sin\pi\left(\frac{n}{2}\right)}{\pi n}\, e^{jn\alpha z^2}$$

This is the Fourier series for zoneplate.

8.10

(a) $$T(x,y) = A_0 + \sum_{n=1}^{\infty} A_n \cos\left[\frac{n\pi}{\lambda R}(x^2+y^2)\right]$$

$$= A_0 + \sum_{n=1}^{\infty} \frac{A_n}{2}\left\{\exp\left[\frac{n\pi}{\lambda R}(x^2+y^2)\right] + \exp\left[\frac{-n\pi}{\lambda R}(x^2+y^2)\right]\right]$$

The zoneplate is illuminated by monochromatic plane wave. So the light field immediately behind the zoneplate is

$$U(x,y) = CA_0 + \sum_{n=-\infty}^{+\infty} \frac{CA_n}{2} \exp\left[\frac{n\pi}{\lambda R}(x^2+y^2)\right]$$

Compare it with a spherical wave

$$U_l(x,y) = e^{-\frac{k}{2l}(x^2+y^2)}$$

, l is the radius of this convergent wavefront,

when $\frac{n\pi}{\lambda R} = \frac{k}{2l} = \frac{2\pi}{\lambda \cdot 2l} = \frac{\pi}{\lambda l}$

or $l = \frac{R}{n}$, we obtain focal point.

so the focal distances of the given zoneplate are $f_n = \frac{R}{n}$, $n = 1, 2, 3, \cdots$

(b)

8.11

(a) $T(x,y)$ may be rewritten as

$$T(x,y) = K_1 + K_2 \cos\left[\frac{2\pi}{\lambda}\left(\frac{x^2 - 2Rx\sin\theta}{2R} + \frac{y^2}{2R}\right)\right]$$

$$= K_1 + K_2 \cos[k(\frac{x^2 - 2Rx\sin\theta + R^2\sin^2\theta - R^2\sin^2\theta}{2R} + \frac{y^2}{2R})]$$

$$= K_1 + K_2 \cos\{k[\frac{(x - R\sin\theta)^2 + y^2}{2R} - \frac{(R\sin\theta)^2}{2R}]\} = K_1 + K_2\cos\phi$$

$$= K_1 + \frac{K_2}{2} e^{j\phi} + \frac{K_2}{2} e^{-j\phi} = T_0 + T_1 + T_2$$

where $\phi = k[\frac{(x - R\sin\theta)^2 + y^2}{2R} - \frac{(R\sin\theta)^2}{2R}]$

When it is illuminated by a monochromatic plane wave, the light field immediately behind the zoneplate is

$U(x,y) = A\,T(x,y) = U_0 + U_1 + U_2$,

where $U_2(x,y) = C \cdot e^{-j\frac{k}{2R}[(x - R\sin\theta)^2 + y^2] - j\phi_0}$

compare it with a spherical wave with a focal point at $(x_0, 0, f)$

$$V(x,y) = C\, e^{-j\frac{k}{2f}[(x - x_0)^2 + y^2]}$$

We can see the focal point of U_2 is located at $(R\sin\theta, 0, R)$

(b)

This zoneplate is called as off-axis zoneplate.

(a) We consider a monochromatic component of the white light first, its wavelength is λ_1. The light field just behind the zoneplate is

$$u(x,y) = A\,T(x,y) = u_0 + u_1 + u_2$$

where $u_2(xy) = C\, e^{-j\frac{k}{2R}[(x-R\sin\theta)^2+y^2]-j\phi_0}$

compare u_2 with a spherical wave of λ_1

$$V(x,y) = C\, e^{-j\frac{k_1}{2f_1}[(x-x_0)^2+y^2]}$$

Then x_0 still equals to $R\sin\theta$,

but $f_1 = \frac{k_1}{k}R = \frac{\lambda}{\lambda_1}R$,

the longer wavelength, the shorter focal length. The focal point is smeared into rainbow color.

(b)

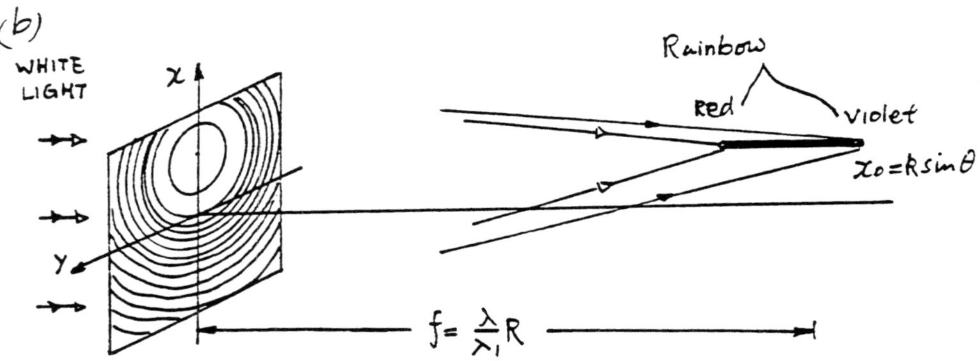

8.13

$$I_c = |u_1 + u_2|^2 = (u_1+u_2)(u_1+u_2)^* = (u_1+u_2)(u_1^* + u_2^*)$$
$$= u_1u_1^* + u_2u_2^* + u_1u_2^* + u_1^*u_2 = |u_1|^2 + |u_2|^2 + u_1u_2^* + u_1^*u_2$$

According to Eq (8.36)

$$I_c = I_1 + I_2 + 2(I_1I_2)^{1/2}\,|\gamma_{12}|\cos\phi_{12}$$

where γ_{12} is the degree of coherence

ϕ_{12} is the phase difference between two light fields

$$\phi_{12} = kx\sin\beta - kx\sin(-\alpha)$$
$$= kx\sin\beta + kx\sin\alpha$$

So the resultant light intensity distribution is $I_c = I_1 + I_2 + 2(I_1I_2)^{1/2}\,|\gamma_{12}|\cos[kx(\sin\alpha + \sin\beta)]$

8.14

(a) Mutually incoherence means $|\gamma_{12}| = 0$

So $I_{inc} = I_1 + I_2 + 0 = I_1 + I_2$

(b) Under coherent case, we see the fringes, Under incoherent case, we cannot see the fringes

coherent

incoherent

8.15

(a) According to Eq. (8.27) the visibility is defined as $V = \frac{I_{Max} - I_{min}}{I_{Max} + I_{min}} = \frac{6-2}{6+2} = \frac{4}{8} = 0.5$

Since two plane waves have equal intensity the degree of coherence

$|\gamma_{12}| = V = 0.5$ see Eq. (8.39)

(b) Refer to Eq. (8.38)

$I_p = 2I[1 + |\gamma_{12}|\cos\phi_{12}]$

$\phi_{12} = kx\sin 45° - kx\sin(-45°) = 2kx\sin 45°$

$I_p = 2I[1 + |\gamma_{12}|\cos(2kx\sin 45°)]$

$= 2I[1 + |\gamma_{12}|\cos(2\pi x \frac{2}{\lambda}\sin 45°)]$

So the spatial frequency

$f_x = \frac{2}{\lambda}\sin 45° = \frac{2 \times \frac{\sqrt{2}}{2}}{500\ nm} = 2828$ lines/mm

8.16

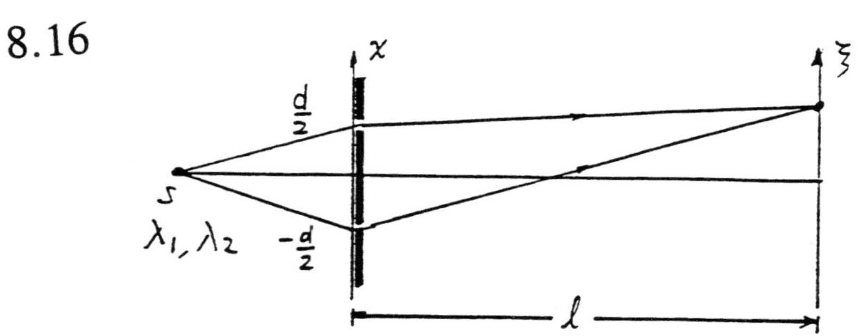

We may draw a system analog diagram:

$$h_\ell = e^{jK\frac{(\xi^2+\eta^2)}{2\ell}}$$

$$u_1 = \delta(x-\tfrac{d}{2}, y) * e^{jK_1\frac{(\xi^2+\eta^2)}{2\ell}} = e^{\frac{jK_1}{2\ell}[(\xi-\frac{d}{2})^2+\eta^2]} = e^{j\phi_1}$$

$$v_1 = e^{j\frac{K_1}{2\ell}[(\xi+\frac{d}{2})^2+\eta^2]} = e^{j\theta_1}$$

$$I_1 = |u_1|^2 + |v_1|^2 + u_1 v_1^* + u_1^* v_1 = 1+1+e^{j\phi_1}e^{-j\theta_1} + e^{-j\phi_1}e^{j\theta_1}$$

$$= 2+2\cos(\phi_1-\theta_1) = 2+2\cos\left[\frac{k_1}{2\ell}(\xi^2+\frac{d^2}{4}+\eta^2-\xi d) - \frac{k_1}{2\ell}(\xi^2+\frac{d^2}{4}+\eta^2+\xi d)\right]$$

$$= 2+2\cos\left[\frac{k_1(2\xi d)}{2\ell}\right] = 2+2\cos\left(2\pi\xi\frac{d}{\lambda_1\ell}\right)$$

Similarly, $I_2 = 2+2\cos\left(2\pi\xi\frac{d}{\lambda_2\ell}\right)$

The resultant intensity $I = I_1 + I_2 = 4 + 2\left[\cos\left(2\pi\xi\frac{d}{\lambda_1\ell}\right) + \cos\left(2\pi\xi\frac{d}{\lambda_2\ell}\right)\right]$

$$= 4 + 4\cos 2\pi\xi\frac{d}{2\ell}\left(\frac{1}{\lambda_1}+\frac{1}{\lambda_2}\right)\cos 2\pi\xi\frac{d}{2\ell}\left(\frac{1}{\lambda_1}-\frac{1}{\lambda_2}\right)$$

Since $\lambda_1 \approx \lambda_2$, $\frac{1}{\lambda_1}+\frac{1}{\lambda_2} \approx \frac{2}{\lambda}$, $\frac{1}{\lambda_1}-\frac{1}{\lambda_2} \approx \frac{\Delta\lambda}{\lambda^2}$, $\lambda = \frac{\lambda_1+\lambda_2}{2}$

$$I \approx 4 + 4\cos\left(2\pi\xi \frac{d}{\ell\lambda}\right)\cos\left(2\pi\xi \frac{d}{\ell}\frac{\Delta\lambda}{2\lambda^2}\right)$$

$$\approx 4 + 4A(\xi)\cos\left(2\pi\xi \frac{d}{\ell\lambda}\right), \quad A(\xi) = \cos\left(2\pi\xi \frac{d}{\ell}\frac{\Delta\lambda}{2\lambda^2}\right)$$

$V = \frac{2A(\xi)}{2B}$, where $B=4$, the average intensity

$$V = \left|\cos\left(2\pi\xi \frac{d}{\ell}\frac{\Delta\lambda}{2\lambda^2}\right)\right|$$

Spatial frequency of I $\quad f_I = \frac{d}{\ell\lambda}$

Spatial frequency of V $\quad f_V = \frac{d}{\ell}\frac{\Delta\lambda}{\lambda^2}$

So $\frac{f_I}{f_V} = \frac{\lambda}{\Delta\lambda} = \frac{5780}{20} = 289$

8.17

According to Eq (8.50)

$$\Delta Y = \frac{\lambda^2}{\Delta\lambda} = \frac{(4880)^2 \ (\text{Å})^2}{0.02 \quad \text{Å}} = \frac{23814400}{0.02}\ \text{Å}$$

$$= 1190720000\,\text{Å} = 11.9\ \text{cm}$$

8.18

$$D_1 = (1.5 - 1)/0.5 = 1$$
$$D_2 = (1.5 - 1)/0.25 = 2$$
$$A = \begin{bmatrix} 1 - 2(0.3)/1.5 & -1 + 2(1)(0.3)/(1.5 - 2) \\ 0.3/1.5 & -1(0.3)/1.5 + 1 \end{bmatrix}$$
$$= \begin{bmatrix} 0.6 & -2.6 \\ 0.2 & 0.8 \end{bmatrix}$$
$$|A| = 1$$

8.19

$$\eta_{t2} = \eta_{air} = 1.0, \quad \eta_{i2} = 1.5$$
$$D_2 = \frac{\eta_{t1}}{d_{21}}(1 - a_{11}) = \frac{1.5}{0.3}(1 - 0.6) = 2$$
$$R_2 = \frac{\eta_{t2} - \eta_{i2}}{2} = \frac{1 - 1.5}{2} = -0.25 \quad (cm.)$$

8.20

$$A = \begin{bmatrix} 1 & -0.05 \\ \frac{1}{1.5} & -\frac{0.05}{1.5} + 1 \end{bmatrix}$$
$$\begin{bmatrix} 0 \\ 2 \end{bmatrix} = A \begin{bmatrix} \alpha_{i1} \\ 2 \end{bmatrix}$$
$$\therefore \alpha_{i1} = 0.1 \quad (rad.)$$

Chapter 9

9.1 Let $\Delta\omega = \omega_1 - \omega_2$, $\Delta\omega \ll \omega_1, \omega_2$

From (9.4)

$$I = \langle A_1^2 \cos^2(\omega_1 t + \alpha_1) + A_2^2 \cos(\omega_2 t + \alpha_2)$$

$$+ A_1 A_2 \cos[(\omega_1 + \omega_2)t + (\alpha_1 \alpha_2)] + A_1 A_2 \cos[\Delta\omega t + \Delta\alpha] \rangle$$

$$= \frac{1}{2} A_1^2 + \frac{1}{2} A_2^2 + A_1 A_2 \langle \cos(\Delta\omega t + \Delta\alpha) \rangle$$

if $\Delta\omega \gg \frac{1}{T}$, T is response time of detector.

then a beat is obtained:

$$I = I_1 + I_2 + \sqrt{I_1 I_2} \cos(\Delta\omega t + \Delta\alpha)$$

9.2 $\delta = \frac{4\pi}{\lambda} \cos\alpha \, (x + y) \stackrel{\text{Let}}{=\!=\!=} 2m\pi$, $m = 0, \pm 1, \pm 2 \cdots$

$$(x + y) = \frac{\lambda}{2\cos\alpha}$$

$$\Lambda = \frac{\sqrt{2}\,\lambda}{4\cos\alpha}$$

9.3 $|E_1 + E_2|^2 = A^2 [\cos^2(\omega t + \frac{2\pi}{\lambda} z) + \sin^2(\omega t + \frac{2\pi}{\lambda} z)]$

$$I = A^2 = 2I_1 = 2I_2 = \text{constant}$$

The output is actually a circularly polarized light.

9.4 $\Lambda = \frac{9.34}{14} = 0.667$ mm

$$d = \frac{\lambda D}{\Lambda} = 949 \ \mu m$$

9.5 (a) $\Delta = 4.5\ \lambda = 2.315\ \mu m$

(b). $(\eta - 1)d = \Delta = 4.5\lambda$

$$d = \frac{\Delta}{(\eta - 1)} = 4.823\ \mu m$$

9.6 maxima position: $x_1 = \frac{m\lambda_1 D}{d}$, $x_2 = \frac{n\lambda_2 D}{d}$

Let $x_1 = x_2 \Rightarrow m\lambda_1 = n\lambda_2$

when $m = 18N$

$n = 21N$, $N = 1, 2, 3 \cdots$

maxima coincide.

9.7 $\delta = \frac{2\pi d x}{\lambda D}$, $d = 2Vt$

$\delta = \frac{4\pi x V t}{\lambda D}$, $I = 4I_1 \cos\left(\frac{2\pi x V t}{\lambda D}\right)$

fringes moving toward the center of screen.

maxima : $x = \frac{m\lambda D}{d} = \frac{m\lambda D}{2Vt}$

$$\frac{dx}{dt} = -\frac{m\lambda D}{Vt^2}$$

9.8

I_1/I_2	1	$\frac{1}{2}$	$\frac{1}{3}$	$\frac{1}{5}$	$\frac{1}{10}$	$\frac{1}{100}$
V	1.0	0.943	0.843	0.745	0.575	0.198

9.9 $\Delta r = \frac{\lambda^2}{\Delta\lambda} = \frac{0.55^2}{0.2} = 1.5125\ \mu m$

$\frac{2\Delta r}{\lambda} = 5.5$

$\therefore$ about $5 \sim 6$ fringes can be seen.

9.10 $\lambda = 0.6328\ \mu m$

$v = 1\ mm/sec = \frac{\lambda}{2} \cdot \frac{N}{t}$

$\frac{N}{t} = \frac{2v}{\lambda} = 3.16\ kHz.$

9.11 (a). $\Delta = 2d\cos\theta$

$\theta = 0, \Rightarrow \Delta = 2d = m_0\lambda$

$m_0 = \frac{2d}{\lambda} = 3000$

10-th fringes: $m = m_0 - 10 = 2990$

$\theta_{10} = \cos^{-1}\left(\frac{m\lambda}{2d}\right) = 4.68°$

(b). similarly, $\theta_{10} = 1.48°$

$m_0 = 30000.$

9.12 $\Delta = 2d = 2 \times 247\ \mu m = 853\lambda$

$\lambda = 0.5791\ \mu m$

9.13 Let $2d = (m+\frac{1}{2})\lambda_1 = n\lambda_2$

$\lambda_1 = 589.6$ nm
$\lambda_2 = 589.0$ nm $\Rightarrow$ $m = 1472.$
$n = 1474$

$2d = n\lambda_2 = 868.2\ \mu m$

9.14 $\eta_1 = 1.000292$

$(\eta_2 - 1) = \frac{P_2}{P_1}(\eta_1 - 1) = 0.0003074$

$N = \frac{(\eta_2 - \eta_1)\,l}{\lambda} = 24.3$

9.15 (a) $A = 1\ m^2$, $\lambda = 0.6328\ \mu m$. $N=1$

$\Omega = \frac{c\lambda N}{4A} = 47.46$ rad/sec.

(b). $A = 1000 \times 1\ m^2$

$\Omega = \frac{c\lambda N}{4A} = 0.04746$ rad/sec

9.16 $\delta = \frac{4\pi}{\lambda}(R - \sqrt{R^2 - r^2}) + \pi \ \underline{\underline{\text{Let}}}\ 3\pi$

$r^2 = R^2 - (R - \frac{\lambda}{2})^2$

$r = 0.287$ mm

9.17 $\delta = \frac{4\pi}{\lambda}\left[R_1 - \sqrt{R_1^2 - r^2} - R_2 + \sqrt{R_2^2 - r^2}\right] + \pi$

$$\approx \frac{4\pi}{\lambda}\left[\frac{r^2}{2R_1^2} + \frac{r^2}{2R_2^2}\right] + \pi \overset{\text{Let}}{=\!=\!=} 3\pi$$

$$r^2 = \frac{\lambda R_1^2 R_2^2}{R_1^2 + R_2^2}, \qquad r = 2.47 \text{ mm}$$

9.18 $\delta = \frac{4\pi n}{\lambda} d\cos\theta + \pi = 2m\pi$

(a). $\theta = 0$, $\frac{4\pi n}{\lambda} d = (2m-1)\pi$

$2m - 1 = 287$, $m = 144$

(b). $\cos\theta = \frac{\lambda}{4nd}(2m-1)$

1st : $m = 143$, $\theta = 6.758°$

2nd : $m = 142$, $\theta = 9.570°$

3rd : $m = 141$, $\theta = 11.731°$

9.19 (a). $R = r^2 = 0.8$,

$$F = \frac{4R}{(1-R)^2} = 80$$

(b). $\Delta\delta = \frac{2(1-R)}{R} = 0.5$

(c). $S = \frac{\pi\sqrt{F}}{2} = 14$

9.20 $R = 0.9 \Rightarrow F = \frac{4R}{(1-R)^2} = 360$

$S = \frac{\pi\sqrt{F}}{2} = 29.8$

Let $\Delta m = \frac{2}{S} = 0.067$

$\delta_1 = \frac{4\pi d_1}{\lambda_1} + \pi = 2m\pi$

$\delta_2 = \frac{4\pi d}{\lambda_2} + \pi = 2(m + \Delta m)\pi$

$d = \frac{\lambda^2 \Delta m}{2\Delta\lambda} = 19.42\ \mu m$

9.21 $\frac{4\pi n d}{\lambda} = (2m+1)\pi$

when $m = 2$, $\lambda = 0.6\ \mu m$.

$m = 3$, $\lambda = 0.428\ \mu m$.

9.22 (a) $n = \sqrt{n_0 n_g} = 1.25$

(b). $nd = \frac{\lambda}{4}$, $d = 0.103\ \mu m$.

9.23 (a). $R_{max} = \left(\frac{n_0 n_g - n^2}{n_0 n_g + n^2}\right) = 0.5$

$\therefore n = 3.0$

(b). $d = \frac{\lambda}{4\eta}, \frac{3\lambda}{4\eta}, \frac{5\lambda}{4\eta}$ -----

$= 0.0429\ \mu m,\ 0.1286\ \mu m,\ 0.2144\ \mu m$ ----

(c). $\sin 45° = 3 \cdot \cos\theta$, $\theta = 13.633°$

$\delta = \frac{4\pi\eta}{\lambda} d\cos\theta = 3.053$ rad

$|\cos\delta| = 0.996$

$r_1^2 = 0.25$, $r_2^2 = 0.10156$

From (9.59)

$R = 40\%$

Chapter 10

10.1

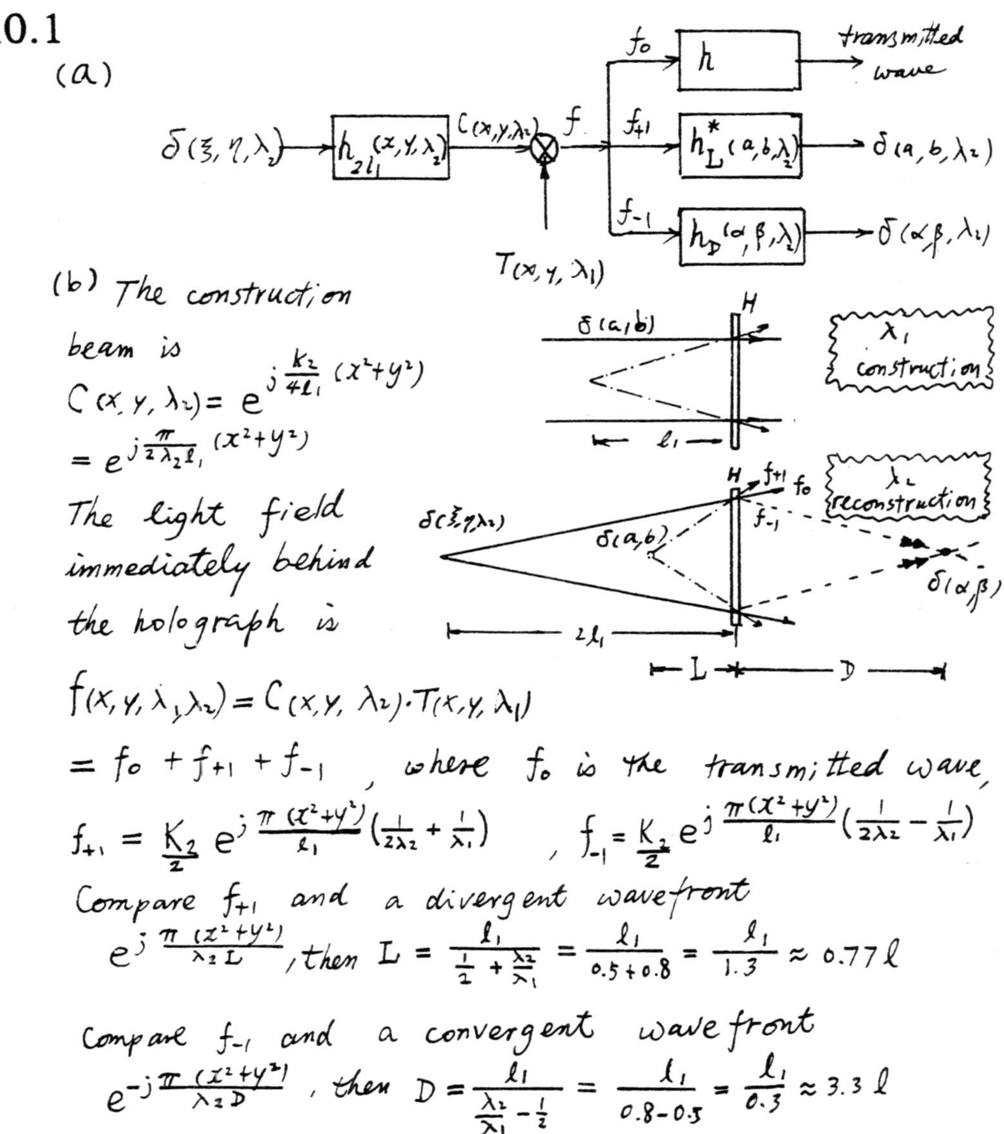

(a)

(b) The construction beam is

$$C(x,y,\lambda_2) = e^{j\frac{k_2}{4l_1}(x^2+y^2)} = e^{j\frac{\pi}{2\lambda_2 l_1}(x^2+y^2)}$$

The light field immediately behind the holograph is

$$f(x,y,\lambda_1,\lambda_2) = C(x,y,\lambda_2)\cdot T(x,y,\lambda_1)$$

$= f_0 + f_{+1} + f_{-1}$, where f_0 is the transmitted wave,

$$f_{+1} = \frac{K_2}{2} e^{j\frac{\pi(x^2+y^2)}{l_1}\left(\frac{1}{2\lambda_2}+\frac{1}{\lambda_1}\right)}, \quad f_{-1} = \frac{K_2}{2} e^{j\frac{\pi(x^2+y^2)}{l_1}\left(\frac{1}{2\lambda_2}-\frac{1}{\lambda_1}\right)}$$

Compare f_{+1} and a divergent wavefront

$$e^{j\frac{\pi(x^2+y^2)}{\lambda_2 L}}, \text{ then } L = \frac{l_1}{\frac{1}{2}+\frac{\lambda_2}{\lambda_1}} = \frac{l_1}{0.5+0.8} = \frac{l_1}{1.3} \approx 0.77\,l$$

Compare f_{-1} and a convergent wave front

$$e^{-j\frac{\pi(x^2+y^2)}{\lambda_2 D}}, \text{ then } D = \frac{l_1}{\frac{\lambda_2}{\lambda_1}-\frac{1}{2}} = \frac{l_1}{0.8-0.5} = \frac{l_1}{0.3} \approx 3.3\,l$$

10.2

(a)

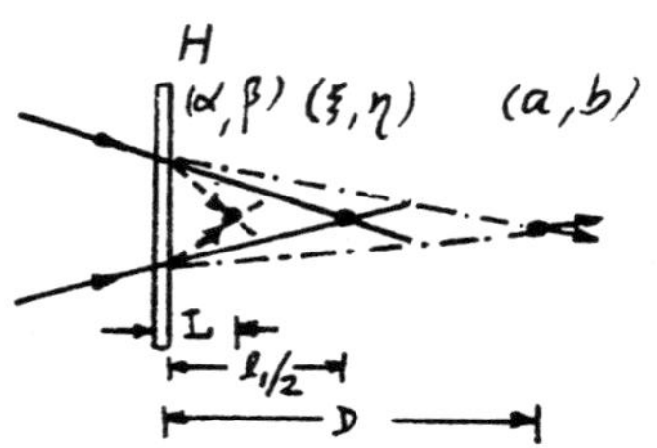

(b)

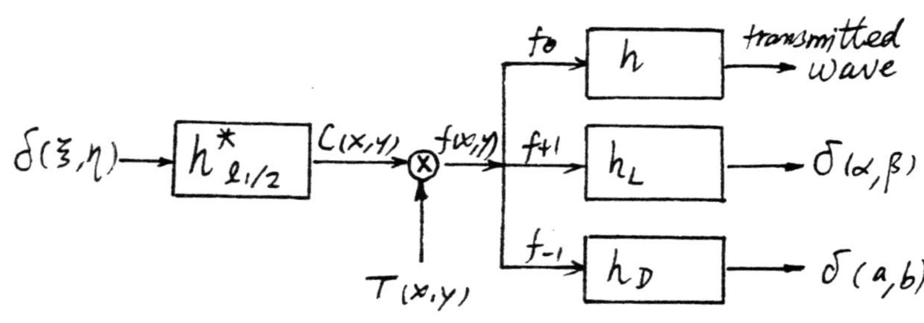

(c) The light field behind the hologram is now

$$f(x,y) = C(x,y) \cdot T(x,y) = f_0 + f_{+1} + f_{-1}$$

where f_0 is the transmitted wave, con e ges at the distance $\frac{1}{2}\ell_1$. $f_1 = K_2 e^{j\frac{\pi(x^2+y^2)}{\lambda_1}\left(-\frac{2}{\ell_1} - \frac{1}{\ell_1}\right)}$

Compare f_1 with a convergent wavefront $e^{-j\frac{\pi(x^2+y^2)}{L}}$, then $L = \frac{1}{3}\ell_1$;

$f_2 = K_2 e^{j\frac{\pi(x^2+y^2)}{\lambda_1}\left(-\frac{2}{\ell_1} + \frac{1}{\ell_1}\right)}$, compare f_2 with another convergent wavefront $e^{-j\frac{\pi(x^2+y^2)}{D}}$. then $D = \ell_1$

10.3

(a) According to Eq (8.4)

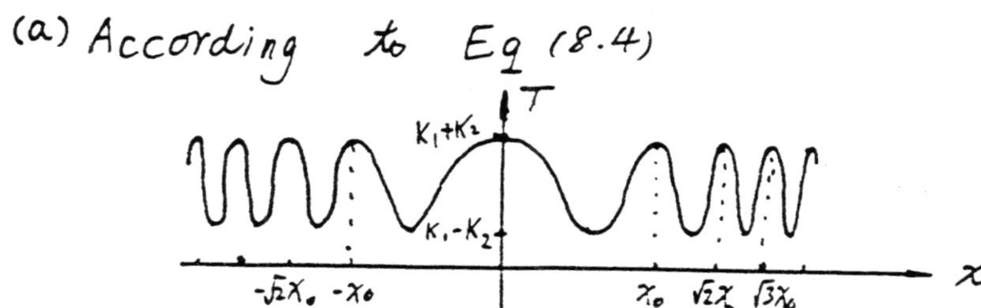

$x_0 = \sqrt{2\lambda_1 \ell_1}$

(b) Near the edge of the hologram, the period $\Delta = \frac{1}{f_x}$, on the otherhand,

$$\frac{\pi (x+\Delta)^2}{\lambda_1 \ell_1} - \frac{\pi x^2}{\lambda_1 \ell_1} = 2\pi \quad , \text{ or } \quad x \cdot \Delta = \lambda_1 \ell_1 .$$

So $f_x = \frac{1}{\Delta} = \frac{x}{\lambda_1 \ell_1} = \frac{5 \text{ cm}}{500 \text{nm} \times 20 \text{ cm}} = 500$ lines/mm

10.4

λ₀ construction

$$T = K_1 + K_2 \cos\left[\frac{\pi}{\lambda_0 l_0}(x^2+y^2)\right]$$

$\lambda_0 = 600$ nm, $l_0 = 50$ cm.

The illuminating wave is

$C(x,y,\lambda) = e^{j\frac{2\pi}{\lambda} x \sin\theta}$

$\leftarrow l_0 \rightarrow$

The light field behind the hologram is

$g(x,y,\lambda) = T(x,y,\lambda_0) \cdot C(x,y,\lambda) = g_0 + g_1 + g_{-1}$

We only consider g_{-1}.

$$g_1(x, y, \lambda_0, \lambda) = e^{-j\frac{\pi}{\lambda_0 l_0}(x^2+y^2) + \frac{2\pi}{\lambda} x \sin\theta}$$

Compare g_1 and a convergent wave $e^{-j\frac{\pi}{\lambda l}[(x-a)^2+y^2]}$

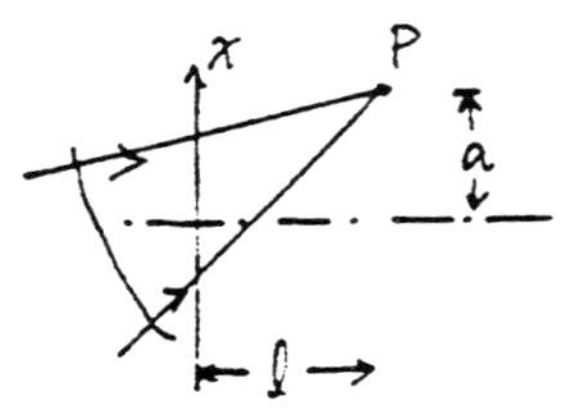

$$= e^{-j\frac{\pi}{\lambda l}[x^2 - 2ax + y^2]} + K_3$$

so $\frac{1}{\lambda_0 l_0} = \frac{1}{\lambda l}$, or $l = \frac{\lambda_0}{\lambda} l_0$, a similar result as Eq (10.15);

and $a = l \sin\theta$.

Corresponding to

$\lambda_1 = 350$ nm, we get

$l_1 = \frac{600}{350} l_0 = 85.7$ cm

$a_1 = l_1 \sin\theta = 60.6$ cm

P_1 P_0 P_2 θ

$\lambda_2, \lambda_0, \lambda_1$ reconstruction

$\lambda_2 = 700$ nm

$l_2 = \frac{600}{700} l_0 = 42.9$ cm

$a_2 = 30.3$ cm.

But in reality, $a_1 = l_1$, $a_2 = l_2$, for the paraxial approximation is unvalid when $\theta = 45°$.

The smeared images cover a length of $\sqrt{(l_2 - l_1)^2 + (a_2 - a_1)^2} = 52$ cm, (actually 60.6 cm)

10.5

(a) Referring to Example 9.1., the image should be located at $l = \frac{\lambda_0}{\lambda} l_0$, for red light

$l_r = \frac{\lambda_0}{\lambda_r} l_0 = \frac{600}{700} \times 0.2 = 0.17m$

for violet light $l_v = \frac{\lambda_0}{\lambda_v} l_0 = \frac{600}{350} \times 0.2 = 0.34m$

(b) the smearing length $\Delta l = l_v - l_r = 0.17m$

(c) Since the smearing length $\Delta l = \lambda_0 l_0 \left(\frac{1}{\lambda_r} - \frac{1}{\lambda_v}\right)$ the color smear will be less when the point object goes closer to the holographic plane during the construction. If $l_0 = 0$, then no smear at all.

10.6

Refer to Eq. (10-3). The maximum and the minimum light intensity on the hologram is $I_M = |C|^2 + |C'|^2 + 2|C||C'|$ and

$I_m = |C|^2 + |C'|^2 - 2|C||C'|$, respectively.

So the modulation index is

$$m = \frac{I_M - I_m}{I_M + I_m} = \frac{2|C||C'|}{|C|^2 + |C'|^2} = \frac{2\sqrt{I/I'}}{1 + I/I'} = \frac{2\sqrt{1/3}}{1 + 1/3} = 0.866$$

Under best recording condition, the modulation

index of the transmittance function is also m, for the recording is in linear region of $t-E$ curve, then

$$T = \frac{1}{2} + (0.866/2)\cos\left[\frac{\pi}{\lambda_1 \ell_1}(x^2+y^2)\right]$$

When T is illuminated by a monochromatic plane wave K, the light field behind T is

$$g = g_0 + g_{+1} + g_{-1}$$

$$g_0 = \frac{1}{2}K, \qquad g_{+1} = \frac{m}{4}K e^{j\phi}, \quad g_{-1} = \frac{m}{4}K e^{-j\phi},$$

Thus, the diffraction efficiency is

$$D.E. = \frac{(g_{+1})^2}{K^2} = \left(\frac{m}{4}\right)^2 = \frac{0.75}{16} = 4.7\%$$

10.7

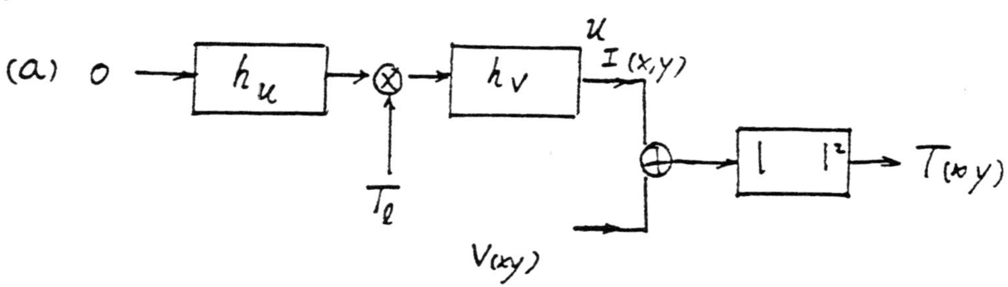

(b) The image is composed by numerous points. Suppose P_i, the image of a point P_o of the object is located at $(A, -l)$. After approximation, the wavefront from P_i is $u = e^{j\frac{k(x-A)^2+y^2}{2l}}$

the reference wave

$v = e^{jkx\sin\theta}$

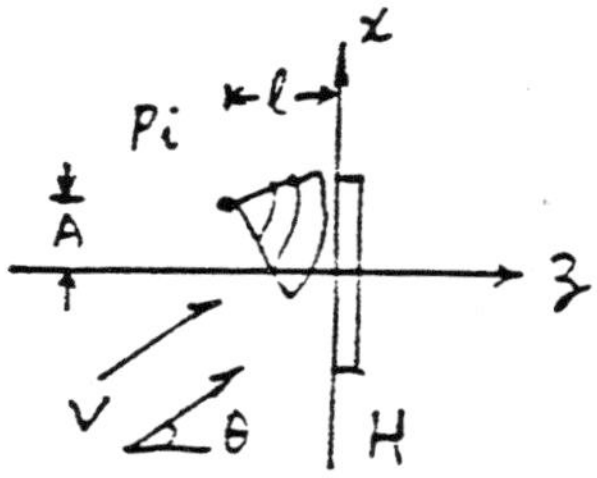

The recorded transmittance

is $T(x, y, \lambda)$

$$= K_1 + K_2 e^{j(\phi_u - \phi_v)} + K_2 e^{j(\phi_v - \phi_u)}$$

Suppose the reconstructing beam is

$V_1 = e^{jk_1 x\sin\theta_1}$, the reconstructed wave is $g(x, y, \lambda, \lambda_1) = V_1 \cdot T = g_0 + g_1 + g_{-1}$.

we only consider $g_1 = e^{j(\phi_u + \phi_{v_1} - \phi_v)}$

$$= e^{j\left\{\frac{k}{2l}[(x-A)^2 + y^2] + k_1 x\sin\theta_1 - kx\sin\theta\right\}}$$

Compare g_1 and a spherical wave

$e^{j\left\{\frac{k_1}{2l_1}[(x-B)^2 + y^2]\right\}}$ which is from the holographic point image. We see

$$\frac{k_1}{l_1} = \frac{k}{l}, \quad \therefore\ l_1 = \frac{k_1}{k} l = \frac{\lambda}{\lambda_1} l$$

$$B = A - \frac{k_1}{k} l\sin\theta_1 + l\sin\theta = A - \frac{\lambda}{\lambda_1} l\sin\theta_1 + l\sin\theta.$$

We define longitudinal color blur as $l_1 - l = l\left(\frac{\lambda}{\lambda_1} - 1\right)$

and lateral color blur as $B - A = l\left(\sin\theta - \frac{\lambda}{\lambda_1}\sin\theta_1\right)$

when $l \to 0$, both $(l_1 - l)$ and $(B - A) \to 0$.

(c) Many monochromatic holographic images are thus superimposed with zero blure, which means a white light image is reconstructed.

10.8

(a) $\delta(O) \rightarrow$ [h_{l_1}] $\rightarrow u(x,y) \rightarrow \oplus$

$\delta(P) \rightarrow$ [$h^*_{l_2}$] $\rightarrow v(x,y) \rightarrow \oplus$

$\oplus \rightarrow$ [$|\ \ |^2$] $\rightarrow T(x, y, \lambda)$

(b) Suppose the coordinates for O and P are $(x_o, 0, -l_1)$ and $(x_R, 0, l_2)$ respectively.

$$u(x,y) = \delta(O) * e^{j\frac{k}{2l_1}[(x-x_o)^2+y^2]} = O\, e^{j\frac{k}{2l_1}[(x-x_o)^2+y^2]}$$

$$v(x,y) = \delta(P) * e^{-j\frac{k}{2l_2}[(x-x_R)^2+y^2]} = R\, e^{-j\frac{k}{2l_2}[(x-x_R)^2+y^2]}$$

The transmittance function is $T = |u+v|^2$

$$= |O|^2 + |R|^2 + OR\, e^{j(\phi_o-\phi_R)} + OR\, e^{j(\phi_R-\phi_o)}$$

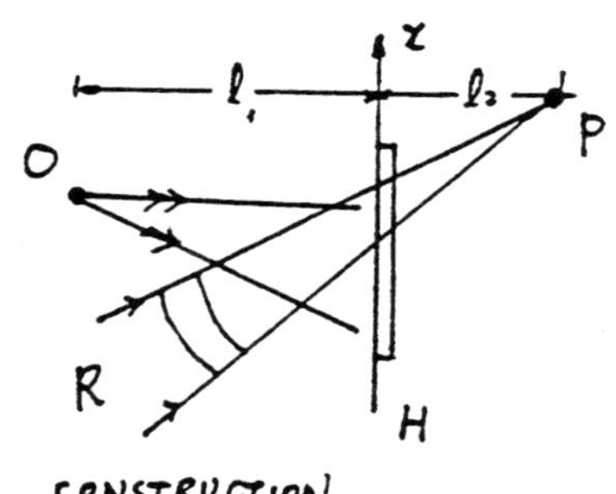

CONSTRUCTION

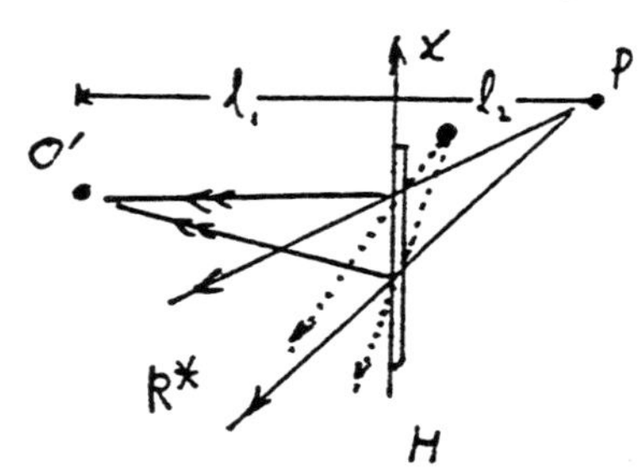

RECONSTRUCTION

10.8

10.9

(a)

$\delta(p) \rightarrow$ [h_{l_2}] $\rightarrow C \rightarrow \otimes \rightarrow g$, with $T(xy)$ entering the multiplier $\otimes$;

$g \rightarrow g_0 \rightarrow$ [transmitted wave] $\rightarrow$

$g \rightarrow g_1 \rightarrow$ [h_{l_1}] $\rightarrow \delta(o')$

$g \rightarrow g_{-1} \rightarrow$ [$h^*_{l_3}$] $\rightarrow \delta(o'')$

(b) The reconstructed wave is composed of three terms, i.e. $g = g_0 + g_1 + g_{-1}$, where

$$g_{-1} = K\, e^{j(\phi_o - \phi_R) - j\phi_R} = K\, e^{j\phi_o - 2\phi_R},$$

$$g_{1} = K\, e^{j(\phi_R - \phi_O) - j\phi_R} = K e^{-j\phi_o} = K\, O^*$$

So there is a real image O' at the original location of the object, $(x_o, 0, -l_1)$.

Compare g_{-1} and a divergent wave.

Suppose $e^{j\frac{k}{2l_1}[(x-x_o)^2+y^2] + 2\frac{jk}{2l_2}[(x-x_R)^2+y^2]}$

$= e^{+j\frac{k}{2L}[(x-A)^2+y^2]}$, by simple calculation, we have $L = \dfrac{+1}{\frac{1}{l_1} + \frac{2}{l_2}}$, and

$$A = +L\left(\frac{x_o}{l_1} + \frac{2x_R}{l_2}\right).$$

10.10

(a) For simplicity, let $x_o = 0$.

$x_R = l_2 \,\mathrm{tg}\, 30° = 11.55$ cm

Suppose the recording wavelength is λ and the reconstructing wavelength is λ_1, we only compare g_{-1} and a convergent wave $e^{i(\phi_R-\phi_o-\phi_{R1})} = e^{-i\phi_{o1}}$, which means $e^{-j\frac{k}{2l_2}[(x-x_R)^2+y^2]-j\frac{k}{2l_1}[x^2+y^2]+j\frac{k_1}{2l_2}[(x-x_R)^2+y^2]}$

$$= e^{-j\frac{k_1}{2L}[(x-A)^2+y^2]}$$

By simple calculation, $\frac{1}{\lambda_1 L} = \frac{1}{\lambda l_1} + \frac{1}{\lambda l_2} - \frac{1}{\lambda_1 l_2}$

so $L = \dfrac{1}{\left(\frac{\lambda_1}{\lambda l_1} + \frac{\lambda_1}{\lambda l_2} - \frac{1}{l_2}\right)}$;

$$A = L\left(\frac{\lambda_1 x_R}{\lambda l_2} - \frac{x_R}{l_2}\right) = \frac{L x_R}{l_2}\left[\frac{\lambda_1}{\lambda} - 1\right]$$

The vertical smearing length is

$$\Delta A = \frac{x_R}{l_2}\left[\left(\frac{\lambda_R}{\lambda}-1\right)L_R - \left(\frac{\lambda_V}{\lambda}-1\right)L_V\right]$$

(b) Substitute the given data into the expressions for L and A,

$L_V = -255$ cm $A_V = 65.8$ cm

$L_R = 23.7$ cm $A_R = 1.5$ cm

$\Delta A = 61.3$ cm

10.11

(a) C → ⊗ → $\mathcal{F}$ → $u(\alpha,\beta)$ → ⊕ → $|\ \ |^2$ → $T(\alpha,\beta)$; $f(x,y)$ → ⊗; $v(\alpha,\beta)$ → ⊕

(b) $u(\alpha,\beta) = \mathcal{F}[f(x,y)] = F(p,q)$

where $p = \frac{2\pi\alpha}{\lambda f}$, $q = \frac{2\pi\beta}{\lambda f}$

$v(\alpha,\beta) = e^{-jk\alpha\sin\theta}$

$T = K_0 + K_1|u|^2 + K_1|v|^2 + K_1 u v^* + K_1 u^* v$

10.12

(a) C → ⊗ → g → $\mathcal{F}^{-1}$ → G_0, G_{+1}, G_{-1}; $T(\alpha,\ )$ → ⊗

(b)

$$
\begin{aligned}
g(x,y) &= C\cdot T(x,y)\\
&= CK_0 + CK_1\left|F\left(\frac{2\pi x}{\lambda f}, \frac{2\pi y}{\lambda f}\right)\right|^2 + K_1 C\\
&\quad + CK_1 F\left(\frac{2\pi\alpha}{\lambda f}, \frac{2\pi\beta}{\lambda f}\right)e^{jk\alpha\sin\theta} + CK_1 F^*\left(\frac{2\pi\alpha}{\lambda f}, \frac{2\pi\beta}{\lambda f}\right)e^{-jk\alpha\sin\theta}\\
&= g_0 + g_{+1} + g_{-1}
\end{aligned}
$$

The output $G = G_0 + G_{+1} + G_{-1}$, where

$$g_0 = \delta(x,y) + f(x,y) \circledast f(x,y)$$

$\circledast$ means auto-correlation

$$g_{+1} = \mathcal{F}^{-1}\left[F(p,q)\,e^{jp f\sin\theta}\right]$$

$$= f(x - f\sin\theta, y)$$

$$g_{-1} = \mathcal{F}^{-1}\left[F^*(p,q)\,e^{-jp f\sin\theta}\right] = f^*(-x - f\sin\theta, -y)$$

(c)

CONSTRUCTION

RECONSTRUCTION

10.13

(a) According to Eq (8.30), the average spatial frequency of the recorded interferogram, i.e. the carrier frequency can determined as follows.

$$T = K_1 + K_2 \cos[\phi(x,y) - k_1 x \sin\theta]$$

We only observe $\cos(k_1 x \sin\theta)$

$= \cos\left(2\pi x \dfrac{\sin\theta}{\lambda_1}\right)$, so the frequency is $f_x = \dfrac{\sin\theta}{\lambda_1} = \dfrac{\sin 45^\circ}{500\ \text{nm}} = 1414$ lines/mm

(b) Eq (8.30) can also be written as

$$T(x,y) = 1 + |u(x,y)|^2 + u(x,y)\, e^{j[k_1 x \sin\theta - \phi(x,y)]} + u(x,y)\, e^{j[\phi(x,y) - k_1 x \sin\theta]},$$

Take its Fourier transform

$$\mathcal{F}[T(x,y)] = \delta(p,q) + \tilde{U}(p,q) \; \tilde{U}(p,q) + \tilde{U}[p+p_0] + \tilde{U}^*(-p+p_0)$$

where $p_0 = \frac{2\pi \sin\theta}{\lambda}$

$\delta(p,q)$

$\tilde{U}(p+p_0)$ $\quad \tilde{U} \circledast \tilde{U}$ $\quad \tilde{U}^*(-p+p_0)$

$-p_0$ $\quad 0$ $\quad P_1$ $\quad P_2$ $\quad P_0$

(c) From the figure, we see $P_1 = 2B(2\pi)$, $P_2 P_0 = B(2\pi)$

$\therefore$ when $\frac{P_0}{2\pi} \geq 3B$, the spectral contents can be separated, i.e. $\frac{\sin\theta}{\lambda} \geq 3 \times 50$ line/mm, $\theta = 4°18'$

10.14

(a) Substitute the given data into Eq (8.50) and Eq (8.51), notice that $L_1 = \infty$,

$$M^r_{Lat} = \left(1 - \frac{\lambda_1 R_1}{\lambda_2 L_2} - \frac{R_1}{\infty}\right)^{-1} = \left(1 - \frac{500\text{ nm} \times 30\text{ cm}}{600\text{ nm} \times 35\text{ cm}}\right)^{-1} = (1 - 0.71)^{-1} = 3.5$$

$$M^{v}_{lat} = \left(1+\frac{\lambda_1 R_1}{\lambda_2 L_2} - \frac{R_1}{\infty}\right)^{-1} = (1+0.71)^{-1} = 0.58$$

(b). When $\lambda_2 = 400$ nm

$$M^{r}_{lat} = \left(1-\frac{\lambda_1 R_1}{\lambda_2 L_2} - \frac{R_1}{\infty}\right)^{-1} = \left(1-\frac{500\times 30}{400\times 35}\right)^{-1} = (1-1.07)^{-1}$$

$$M^{v}_{lat} = \left(1+\frac{\lambda_1 R_1}{\lambda_2 L_2} - \frac{R_1}{\infty}\right)^{-1} = (1+1.07)^{-1} = 0.48$$

(C) From Eq (8.48)

for real images, $\frac{1}{l_r} = \frac{\lambda_2}{\lambda_1 R_1} - \frac{\lambda_2}{\lambda_1 L_1} - \frac{1}{L_2}$

$$l_r = \left(\frac{\lambda_2}{\lambda_1 R_1} - \frac{1}{L_2}\right)^{-1},$$

$\lambda_2 = 600$ nm , $l_r = 87.5$ cm

$\lambda_2 = 400$ nm , $l_r = -116$ cm

Similarly, for virtual images,

$$\frac{1}{l_v} = \frac{\lambda_2}{\lambda_1 L_1} - \frac{\lambda_2}{\lambda_1 R_1} + \frac{1}{L_2}$$

$$l_v = \left(\frac{1}{L_2} - \frac{\lambda_2}{\lambda_1 R_1}\right)^{-1} = -l_r$$

$\lambda_2 = 600$ nm , $l_v = -87.5$ cm

$\lambda_2 = 400$ nm , $l_v = 116$ cm

So the smaller the reconstruction wavelength, the farther the image moves away from the hologram.

10.15

(a). Now substitute $l_2 = -35$ cm instead. $\lambda_2 = 600$ nm

$$M^r_{Lat} = \left[1 - \frac{500\text{ nm} \times 30\text{ cm}}{600\text{ nm} \times (-35\text{ cm})}\right]^{-1} = (1+0.71)^{-1} = 0.58$$

$$M^v_{Lat} = \left[1 + \frac{500\text{ nm} \times 30\text{ cm}}{600\text{ nm} \times (-35\text{ cm})}\right]^{-1} = (1-0.71)^{-1} = 3.5$$

(b). $\lambda_2 = 400$ nm

$$M^r_{lat} = \left[1 - \frac{500 \times 30}{400 \times (-35)}\right]^{-1} = (1+1.07)^{-1} = 0.48$$

$$M^v_{lat} = (1-1.07)^{-1}$$

(c) $\lambda_2 = 600$ nm $\qquad l_r = \left(\frac{\lambda_2}{\lambda_1 R_1} + \frac{1}{35\text{ cm}}\right)^{-1} = 14.6$ cm

$\lambda_2 = 400$ nm $\qquad l_r = 18.1$cm

As for virtual image

$\lambda_2 = 600$ nm $\qquad l_v = \left(-\frac{1}{35\text{ cm}} - \frac{\lambda_2}{\lambda_1 R_1}\right)^{-1} = -14.6$ cm

$\lambda_2 = 400$ nm $\qquad l_v = -18.1$ cm

In problem 14, the images are away from the location of the object, Here on the contrary, the images are located closer to the hologram.

10.16

(a) The wavefront from the object may be expressed as $u(x,y) = O(x,y)\,e^{j\phi_o(x,y)}$.
The reference wave front, $v(x,y) = R(x,y)\,e^{j\phi_R(x,y)}$

The recorded transmittance is

$$T = K I = K|u+v|^2$$

$$= K[\,|O(x,y)|^2 + |R(x,y)|^2 + O(x,y)R(x,y)e^{j(\phi_o - \phi_R)}$$

$$+ O(x,y)R(x,y)e^{j(\phi_R - \phi_o)}]$$

where O and R denoting the amplitude distribution, and ϕ_o, ϕ_R the phase distribution, respectively.

(b) The light field behind the hologram is

$$g(x,y) = K R(x,y) e^{j\phi_R(x,y)} [\,|O(x,y)|^2 + R(x,y)^2]$$

$$+ O(x,y) R^2(x,y) e^{j\phi_o} + O(x,y) R(x,y) e^{j(2\phi_R - \phi_o)}$$

The second term is the regenerated object wavefront, for $R^2(x,y) \approx 1$ (uniform)

(c) Now

$$g'(x,y) = R^* \cdot T = K R(x,y) e^{-j\phi_R(x,y)} [\,|O(x,y)|^2 + |R(x,y)|^2)$$

$$+ O(x,y) R^2(x,y) e^{j(\phi_o - 2\phi_R)} + O(x,y) R^2(x,y)\, e^{-j\phi_o}$$

The third term is the reconstructed conjugate object wavefront.

10.17

(a)

C → ⊗ → h → u → (+) → | |² → T, with V added at (+) and G(x,y) into ⊗

where C is the monochromatic incident wave,

G(x,y) is the complex reflectance.

$G(x,y) = R(x,y)\, e^{j\phi_0(x,y)}$, $R(x,y)$ is the real reflectance; $\phi_0(x,y)$ is the phase factor due to the deviation of the object surface point (x, y) away from the holographic emulsion.

(b) $V = e^{-jk_1 z}$

$u = R(x,y)\, e^{j[\phi_0(x,y)+k_1 z]}$ so the recorded hologram is

$D(x, y, z, k_1) = K_1 + K_2 \cos[k_1(2z + \phi_0(x, y)]$,

(c) When $2k_1 z = 2n\pi$, $n = 0, 1, 2, \ldots$

the interference is constructive,

$\frac{2\pi \times 2z}{\lambda_1} = 2\pi n$, $z = \frac{n\lambda'}{2}$, the separation between subholograms is $\Delta z = \frac{\lambda'}{2}$

10.18

Refer to the solution to problem 8.7. We see both the longitud color blur

$l\left(\frac{\lambda}{\lambda_1}-1\right)$ and lateral color blur $l\left(\sin\theta - \frac{\lambda}{\lambda_1}\sin\theta_1\right)$ will be minimized when $l \to 0$.

10.19

A true color object may be expressed as

$$\mathcal{G}(x,y,\lambda) = G_r(x,y) + G_b(x,y) + G_g(x,y)$$
$$= R_r(x,y)\,e^{j\phi_{or}(x,y)} + R_b(x,y)\,e^{j\phi_b(x,y)} + R_g(x,y)\,e^{j\phi_g(x,y)}$$

Within the linear region of the emulsion, three exposures with different color coherent light may be made one after another, in appropriate time. The recorded hologram may be expressed as

$$\mathcal{T} = K_r t_r |1 + R_r e^{j\phi_r}|^2 + K_b t_b |1 + R_b e^{j\phi_b}|^2 + K_g t_g |1 + R_g e^{j\phi_g}|^2.$$

r = red, b = blue, g = green, t_r means exposure time under red illumination and so on.

Thus a true color reflection hologram is constructed.

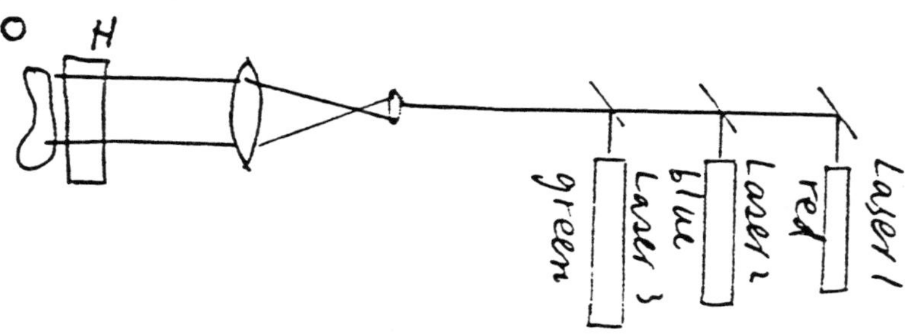

10.20

Refer to the solution to problem 10.16, consider the construction of hologram H_2, $I(x,y)$ is a real image reconstructed from H_1, so the transmittance of H_2 may be expressed as $H_2 = |I + R|^2 = |I|^2 + |R|^2 + IR^* + I^*R$, where $I = O^*(x,y)$, $O(x,y)$ is the original object wave.

Now let us see the reconstruction $C = R^*(x,y)$ so the light field behind H_2 is

$$C \cdot H_2 = R^*[K_1 + IR^* + I^*R]$$
$$= R^*K_1 + IR^*R^* + I^*RR^*$$

the third term is $I^* RR^*$

$= I^* |R|^2 = I^* = (O^*)^* = O$ (say)

so an orthoscopic holographic image of the original object is reconstructed.

10.21

The red image of the slit stays at its original location. As for the violet image, the longitudinal distance is still $l = \frac{\lambda_1}{\lambda} l_1$, so $l_{red} = 30$ cm, $l_{violet} = 40$ cm.

The lateral location of the violet image is now $\alpha_{12} = \pm \frac{W}{2} + \frac{\lambda - \lambda_1}{\lambda} l_1 \sin\theta$

$$= \pm 1 + \frac{400-600}{400} \times 400 \times \sin 50°$$

$$= \pm 1 - 115 \text{ mm}$$

$\alpha_1 = -114$ mm, $\alpha_2 = -116$ mm,

Thus the wavelength spread over the pupil is $\frac{200 \times 2.5}{117} = 4.3$ nm.

which is less than 6.5 nm in Example 10.11.

10.22

$$l = \frac{\lambda_1}{\lambda} l = \frac{600 \text{ nm}}{700 \text{ nm}} \times 30 \text{ cm} = 25.7 \text{ cm}$$

$$\alpha_{1,2} = \pm \frac{W}{2} + \frac{\lambda - \lambda_1}{\lambda} l_1 \sin\theta = \pm 1 + \frac{700-600}{700} \times 300 \times \sin 50^\circ$$

$$= \pm 1 + 32.8 \text{ mm}, \quad \alpha_1 = 31.8 \text{ mm}, \quad \alpha_2 = 33.8 \text{ mm}$$

The results mean that the longer the reconstructing wavelength, the larger the diffraction angle.

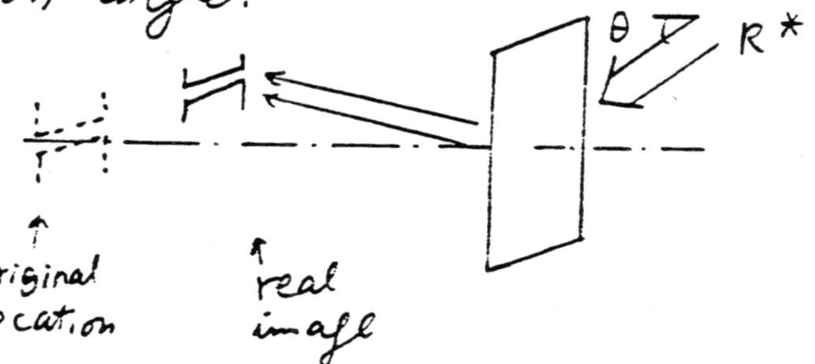

10.23

According to Eq (12.26) in Reference [1] the color blur ΔH_c is proportional to the width of the slit. So the narrower the slit, the less the color blur.

10.24

(a). As an alternative method, here we use geometric optics.

$$\frac{1}{u_s} + \frac{1}{V_s} = \frac{1}{f} \quad , \quad \text{since } u_s = \frac{f}{2}$$

$$V_s = \frac{u_s \cdot f}{u_s - f} = \frac{\frac{f}{2} \cdot f}{\frac{f}{2} - f} = -f$$

$M_{lat} = \left|\frac{V_s}{u_s}\right| = 2$, so the width of the image is $2W$.

(b).

O S F F H u_s u_0 d V_0

S_I R θ x $\tilde{u}_1$ $\tilde{u}_2$ O(x,y) ℓ d α ξ

$\ell = V_0 - d + V_s$

$\tilde{u}_1 = \delta(\alpha - W)$

$\tilde{u}_2 = \delta(\alpha + W)$

$R = e^{-jkx\sin\theta}$

(c)

$\delta(\alpha - W) \rightarrow$ [h_ℓ] $\rightarrow u_1$

$\delta(\alpha + W) \rightarrow$ [h_ℓ] $\rightarrow u_2$

$\delta(\xi) \rightarrow$ [h^*_d] $\rightarrow u$

$R \rightarrow$

$\oplus \rightarrow$ [$|\ |^2$] $\rightarrow T_{RAINBOW}$

10.25

(a) In thick emulsion, there are a lot of sub hologram. The reconstruction must satisfy the Bragg's Law.
So the reconstructing beam is the conjugate of the reference beam.

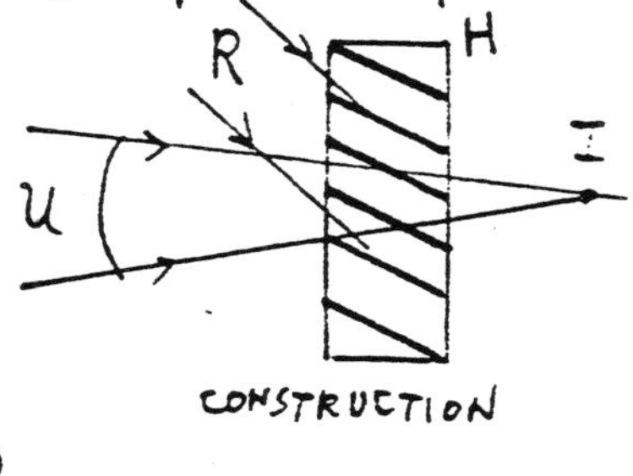

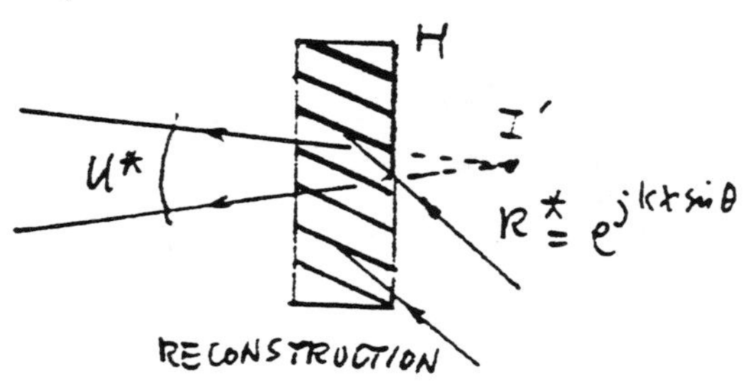

(b)

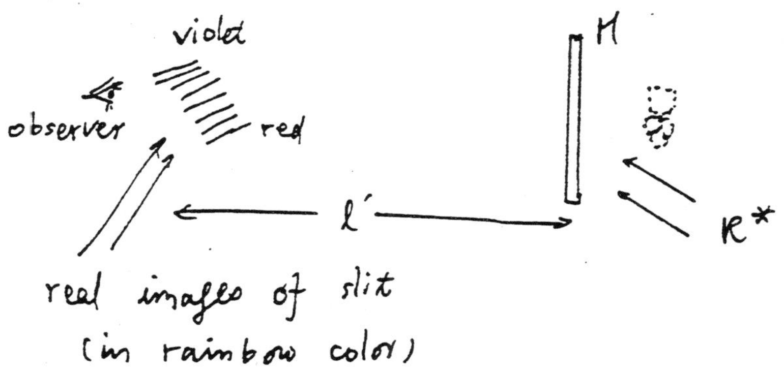

10.26

(a) $\frac{1}{u_s} + \frac{1}{v_s} = \frac{1}{f}$, $v_s = \frac{f u_s}{u_s - f} = \frac{f \times 1.5f}{1.5f - f} = 3f$

$M = -\frac{v_s}{u_s} = -\frac{3f}{1.5f} = -2$, so $W_I = 2w$.

(b) This is a real image of the slit. it is located behind the lens. In problem 10.24, a virtual image of the slit is located before the lens. The lateral magnification in both case is the same, 2, so the width of the images equals to 2W

(c)

S L F F H u_s u_o d V_o

R θ x O(x,y) H d ξ S_I $\tilde{u}_2$ $\tilde{u}_1$ l

$l = V_s - V_o + d$

$\tilde{u}_1 = \delta(\alpha + w)$

$\tilde{u}_2 = \delta(\alpha - w)$

$R = e^{-jkx\sin\theta}$

10.27

(a) point-concept analog system diagram:

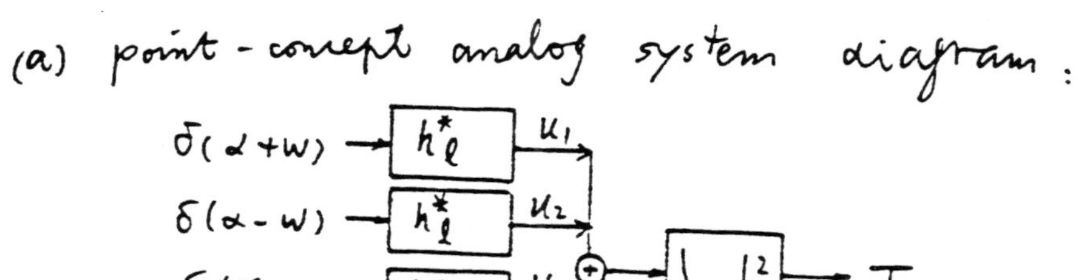

the imaging process: suppose the object is a convex one.

A B C object   lens   H C′ B′ A′ real image   (A)

CONSTRUCTION   R   H C′ B′ A′   (B)

RECONSTRUCTION   C=R   H C″ B″ A″   viewer (C)

(b) As a demonstration, we choose a convex object ABC, apparently $u_A > u_B$, $u_C > u_B$. according to the lens formula $V = \frac{uf}{u-f} = \frac{f}{1-\frac{f}{u}}$

$V_A < V_B$, $V_C < V_B$, in other words, the real image is orthoscopic, refer to (A)

This real image is recorded and reconstructed, remains orthoscopic.

In (B), since $V_A < V_B$, $V_C < V_B$,

$d_A' < d_B'$, $d_C' < d_B'$

According to Eq (10.15), the distance of the reconstructed image from the hologram is

$$d'' = \frac{\lambda_1}{\lambda_2} d'$$

so $d_A'' < d_B''$, $d_C'' < d_B''$. The viewer see an orthoscopic image [see FIG (C)].

10.28 a) $d = 1\,mm$; $\eta = 2.28$; $\lambda = 630\,nm$

Using eq. (10.58) (assuming plane wave object beam) for simplicity (closer approximation can be derived with a method similar to the one described in section 11.7)

$$\left(\frac{\Delta\lambda}{\lambda}\right)_t = \frac{(2.28^2 - \sin^2\alpha)^{1/2}}{\sin^2\alpha} \times 630 \times 10^{-6}$$

b)

$$\frac{d}{\lambda}\left(\frac{\Delta\lambda}{\lambda}\right) = \frac{(\eta^2 - \sin^2\alpha)^{1/2}}{\sin^2\alpha}$$

$$\frac{d}{\lambda}\left(\frac{\Delta\lambda}{\lambda}\right)_t = \underline{\Delta(\alpha)}$$

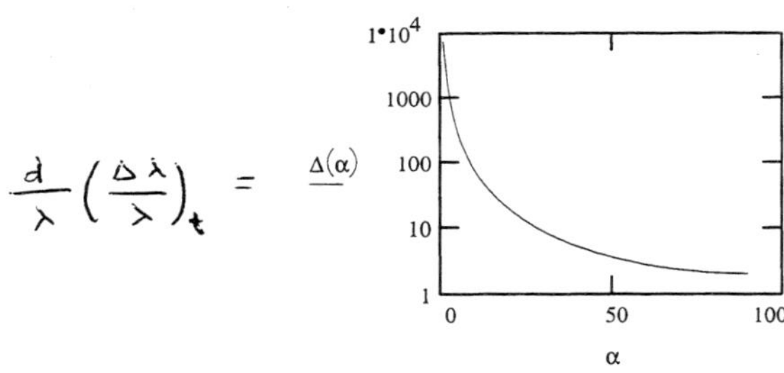

10.29 (a) Again for simplicity we assume a plane wave object beam so that we can use eq. (10.59)

$$\left(\frac{\Delta\lambda}{\lambda}\right)_r = \frac{1}{(\eta^2 - \cos^2\alpha)^{1/2}} \frac{\lambda}{d}$$

$$= \frac{1}{(2.28^2 - \cos^2\alpha)^{1/2}} \cdot 630 \times 10^{-6}$$

$$\Delta_r(\alpha) = \frac{d}{\lambda}\left(\frac{\Delta\lambda}{\lambda}\right)_r = \frac{1}{(2.28^2 - \cos^2\alpha)^{1/2}}$$

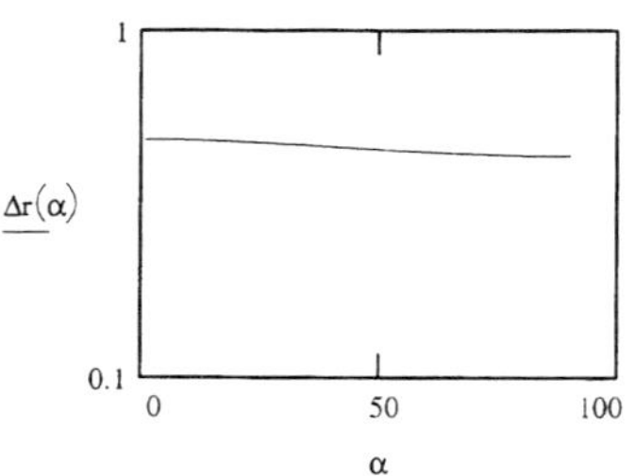

Comparing this to the result of prob 10.28 it is clear that the reflection structure has better selectivity. In addition to that it's value does not change as much when the angle changes.

10.30

(a) Using eq 10.58 as an approximation and

$\lambda = 630$ as a center wavelength; $\eta = 2.28$

$$\Delta\lambda = \frac{(\eta^2 - \sin^2\alpha)^{1/2}}{\sin^2\alpha} \frac{\lambda^2}{d}$$

take $\alpha = 30^\circ$ as an example, and $d = 1$ mm

$$\Delta\lambda = \frac{(2.28^2 - 0.5^2)^{1/2}}{0.5^2} \; 630 \times 10^{-6} \times 630 \text{ nm}$$

$$\simeq 0.397 \text{ nm}$$

if we have a tunable laser diode from $\lambda = 620 - 640$ nm

we can have $\simeq \frac{20}{0.4} = 50$ holograms

(b) The smallest pixel size $\rightarrow \frac{a}{m} \times \frac{a}{m}$

min. period $= \frac{2a}{m} \rightarrow$ max. spatial freq $= \frac{m}{2a}$

using $p = 2\pi f_{sp} = \frac{2\pi}{\lambda} f \alpha$

where f_{sp} is the spatial frequency, $f =$ fourier lens focal length

and α is the spatial frequency coordinate (at P_1 plane)

$\alpha_{max} = \frac{m\lambda}{2a} f \Rightarrow$ the size of the frequency spectrum at

$P_1 = 2 \times \alpha_{max} = \frac{m\lambda f}{a}$ (note: consider both positive and negative α axis)

Now if the collimated beam angle (the one that travels toward L_1) is changed, the center frequency distribution at P_1 will be shifted, therefore by employing $N \times N$ pinhole array at P_1 in association with the different angle of beam impinging L_1, N^2 objects can be multiplexed in the crystal

Chapter 11

11.1 a) According to Eq. (11.15), the output intensity from an incoherent system is

$$I(\alpha,\beta) = |h(x,y)|^2 * |f(x,y)|^2.$$

Take Fourier transform of both sides of this equation. $\mathcal{F}[I(\alpha,\beta)] = \mathcal{F}[|h(x,y)|^2] \cdot \mathcal{F}[|f(x,y)|^2]$.

The incoherent transfer function is then

$H_i(p,q) = \mathcal{F}[|h(x,y)|^2] = H_c(p,q) \circledast H_c(p,q)$, where $H_c(p,q) = \mathcal{F}[h(x,y)]$ is the coherent transfer function of the given system. So the incoherent transfer function is the autocorrelation of the coherent transfer function.

b) ① $\iint_{-\infty}^{+\infty} H_i(p,q)\,dp\,dq = 1$ (normalization condition).

② $H_i(0,0) \geq H_i(p,q)$, the maximum occurs at the origin of the frequency plane.

③ If $H_c(p,q)$ is real, then $H_i(p,q)$ is real and symmetrical.

④ The cutoff frequency of incoherent transfer function is as much as twice the cutoff frequency of coherent transfer function.

For simplicity, we suppose there is only one aberration-free lens in this system.

The impulse response under monochromatic coherent illumination is

P L

point object

image

(α,β)

(x,y)

d_o d_i

$$h_c = \{[\delta(\alpha,\beta) * h_{do}] \cdot T_l\} * h_{di}$$

where

$$T_l = P(\xi,\eta)\, e^{-i\frac{k}{2f}(\xi^2+\eta^2)}$$

$\delta(\alpha,\beta)$ → h_{do} → ⊗ → h_{di} → $h_c(x,y)$; $T_l(\xi,\eta)$ → ⊗

$$h_{do} = \exp[i\frac{k}{2d_o}(\xi^2+\eta^2)]$$

$$h_{di} = \exp[i\frac{k}{2d_i}(x^2+y^2)]$$

$$\therefore h_c = [P(\xi,\eta)\, e^{-i\frac{k}{2}(\frac{1}{f}-\frac{1}{d_o})(\xi^2+\eta^2)}] * e^{i\frac{k}{2d_i}(x^2+y^2)}$$

$$= \iint P(\xi,\eta)\, e^{-i\frac{k}{2}(\frac{1}{f}-\frac{1}{d_o})(\xi^2+\eta^2)}\, e^{i\frac{k}{2d_i}[(x-\xi)^2+(y-\eta)^2]}\, d\xi\, d\eta$$

$$= K \iint P(\xi,\eta)\, e^{-i\frac{k}{2}(\frac{1}{f}-\frac{1}{d_o}-\frac{1}{d_i})(\xi^2+\eta^2)}\, e^{-i\frac{k}{2d_i}(x\xi+y\eta)}\, d\xi\, d\eta$$

Let $\frac{1}{d_i} = \frac{1}{f} - \frac{1}{d_o}$

$h_c(x,y,d_i) = \mathcal{F}[P(\xi,\eta)] = \tilde{P}(p,q)$, where $p = \frac{kx}{d_i}$, $q = \frac{ky}{d_i}$. So the impulse response is the Fourier transform of the aperture function, in other words, its Fraunhofer diffraction pattern. The transfer function is the Fourier

transform, $H_c(f_x, f_y) = \mathcal{F}[h] = \mathcal{F}\{\mathcal{F}[P(\xi,\eta)]\}$
$= P(\lambda d_i f_x, \lambda d_i f_y)$, where $f_x = \frac{p}{2\pi} = \frac{x}{\lambda d_i}$, $f_y = \frac{q}{2\pi} = \frac{y}{\lambda d_i}$. For example, $P(\xi,\eta) = \mathrm{circ}\left(\frac{r}{r_0}\right)$, $r = \sqrt{\xi^2+\eta^2}$, (most apertures are circular), then

$$H_c(f_x,f_y) = P(\lambda d_i f_x, \lambda d_i f_y) = \mathrm{circ}\left(\frac{\rho}{r_0}\right)^2, \quad \rho = \lambda d_i \sqrt{f_x^2+f_y^2}$$

when $\rho = r_0$, $H_c = 0$.

so the cutoff frequency

$$f_c = \frac{r_0}{\lambda d_i}$$

From the solution to the previous problem, we know the incoherent transfer function $H_i = H_c \circledast H_c$

hence the cutoff frequency is $f_i = 2f_c = \frac{2r_0}{\lambda d_i}$

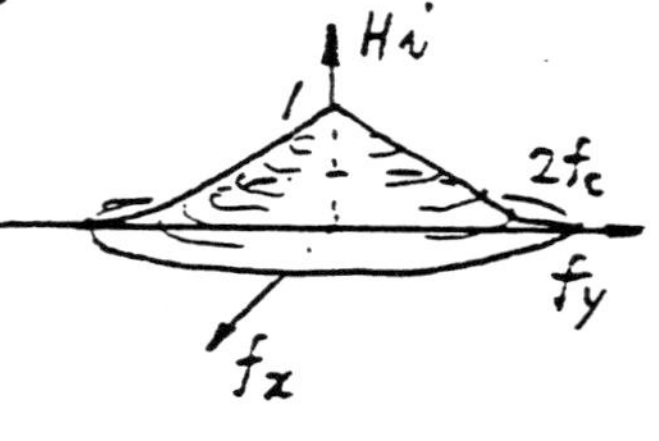

(r_0 is the radius of the aperture)

11.3 Under coherent illumination,

$f_1(x,y) \longrightarrow F_1(\alpha,\beta)$, $f_2(x,y) \longrightarrow F_2(\alpha,\beta)$,

The irradiance is

$$I = |F_1(\alpha,\beta) + F_2(\alpha,\beta)|^2 = (F_1+F_2)(F_1+F_2)^*$$

$= |F_1|^2 + |F_2|^2 + F_1 F_2^* + F_1^* F_2$, where F_1 and F_2 are the diffraction field of f_1 and f_2, respectively. If the illumination beams on f_1 and f_2 are incoherent, then the irradiance

$$I = I_1 + I_2 = |F_1|^2 + |F_2|^2$$

11.4 Coherent illumination: the input is $f(x,y) = e^{i\phi(x)}$

the output field $g(x,y) = f(x,y) * h(x,y)$

$= e^{i\phi(x)} * \delta(x,y) = e^{i\phi(x)}$

the output irradiance $I(x,y) = g\,g^* = 1$

Incoherent illumination: the input is $I_i(x,y) = f f^* = 1$

the output field $I(x,y) = I_i(x,y) * |h(x,y)|^2 = 1 * \delta(x,y) = 1$

But under coherent illumination, the observer will see speckles floating in the air.

11.5 According to Euler's formula, the input

$f(x,y) = \frac{1}{2}[1 + \cos(60\pi x)] = \frac{1}{2} + \frac{1}{4} e^{i2\pi \cdot 30x} + \frac{1}{4} e^{-i2\pi \cdot 30x}$

At the Fourier plane, we obtain

$F(p,q) = \left[\frac{1}{2}\delta(p) + \frac{1}{4}\delta(f_x - f_0) + \frac{1}{4}\delta(f_x + f_0)\right]\delta(f_y)$

where $f_0 = 30\,l/m$, $f_x = \frac{p}{\lambda f}$, $f_y = \frac{q}{\lambda f}$.

a) Under violet illumination, one peak is at the origin of the Fourier plane, other two peaks are located at $p_v = \pm \lambda_v f f_o = \pm 400\,nm \times 300\,mm \times 30\,mm^{-1}$ $= \pm 3.6\,mm$

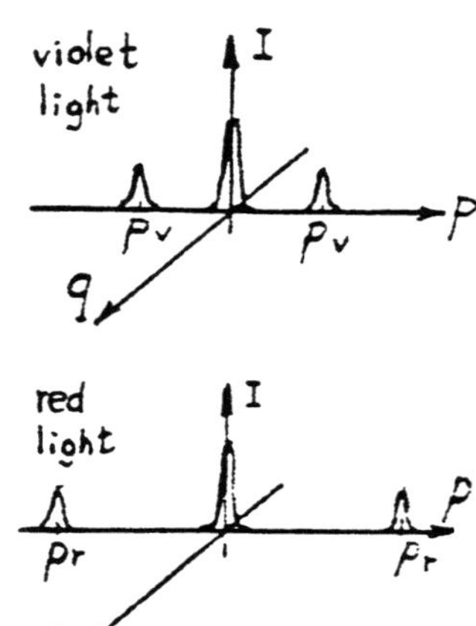

b) under red light illumination. so the main peak stays at the origin, the other two depart from their locations.

$$p_r = \pm \lambda_r f f_o$$
$= \pm 600\,nm \times 300\,mm \times 30\,mm^{-1} = \pm 5.4\,mm$. The change of locations is due to the different wavelengths.

11.6 At the frequency plane, the coming light field is $F(p,q) = [\frac{1}{2}\delta(p) + \frac{1}{4}\delta(f_x - f_o) + \frac{1}{4}\delta(f_x + f_o)]\,\delta(q)$,

the leaving light field is

$$G(p,q) = \frac{1}{4}[\delta(f_x - f_o) + \delta(f_x + f_o)]\,\delta(q)$$

So at the output plane, we get

$$g(\alpha,\beta) = \frac{1}{2}\cos(2\pi f_o \alpha)$$

a) The output intensity

$$I_g(\alpha,\beta) = |g(\alpha,\beta)|^2 = \frac{1}{4}\cos^2(2\pi f_o \alpha)$$
$$= \frac{1}{8}[1 + \cos(2\pi\, 2f_o \alpha)]$$

b) The basic frequency of $I_g(\alpha,\beta)$ is $f' = 2f_o = 60$ line/mm.

11.7 The input object may be expressed as $f(x,y) = \text{rect}\left(\frac{x}{s}, \frac{y}{s}\right)$

(a) Its Fourier spectrum is, according to the scaling property, see §7.5.3.,

$$F(f_x, f_y) = s^2 \,\text{sinc}(s f_x, s f_y),$$

where $f_x = \frac{p}{2\pi} = \frac{\alpha}{\lambda f}$, $f_y = \frac{q}{2\pi} = \frac{\beta}{\lambda f}$,

when sf_x or $sf_y = 1, 2, 3$, $F(f_x, f_y) = 0$

maximum occurs at $f_x = f_y = 0$.

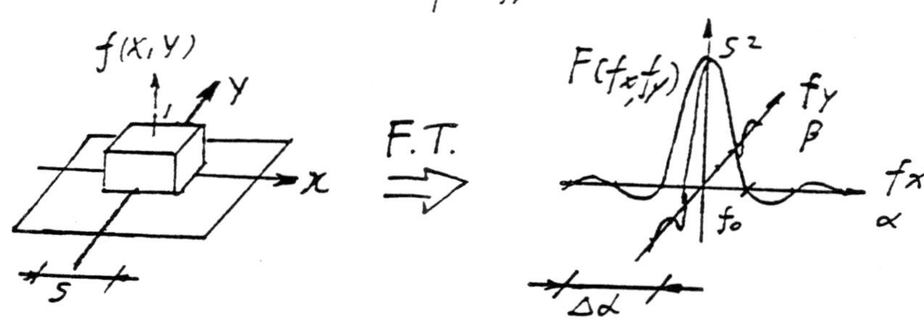

$s f_0 = 1, \quad f_0 = \frac{1}{s}\ ; \qquad \alpha = \lambda f f_x \quad \therefore \alpha_0 = \frac{\lambda f}{s}$

$\Delta\alpha = 2\alpha_0 = \frac{2\lambda f}{s} = \frac{2 \times 600 \times 10^{-6}\,\text{mm} \times 1{,}000\,\text{mm}}{5\,\text{mm}} = 0.24\,\text{mm}$

$\Delta\alpha$ is the width of the main lobe.

(b) when $f_2 = 100$ mm, $\Delta\alpha_2 = \frac{2\lambda f_2}{S} = \frac{2 \times 600 \times 10^{-6} \times 100}{5}$ mm

$= 0.024$ mm

11.8 a) The input light field is $f(x,y) + n(x,y)$, its Fourier transform $F(p,q) + N(p,q)$ appears at plane P_2, immediately behind P_2 is $[F(p,q) + N(p,q)]F^*(p,q)$. So at the second focal plane of the inverse Fourier transform lens L_2, the output light distribution is

$[f(\alpha,\beta) + n(\alpha,\beta)] \circledast f(\alpha,\beta) = f(\alpha,\beta) \circledast f(\alpha,\beta) + \underbrace{n(\alpha,\beta) \circledast f(\alpha,\beta)}_{0}$

$= f(\alpha,\beta) \circledast f(\alpha,\beta)$

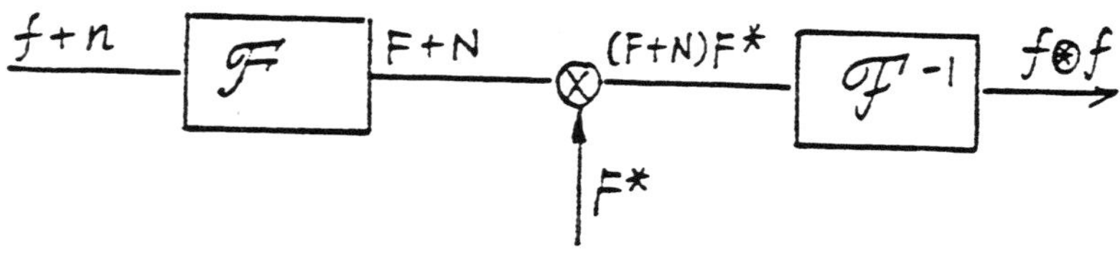

b)

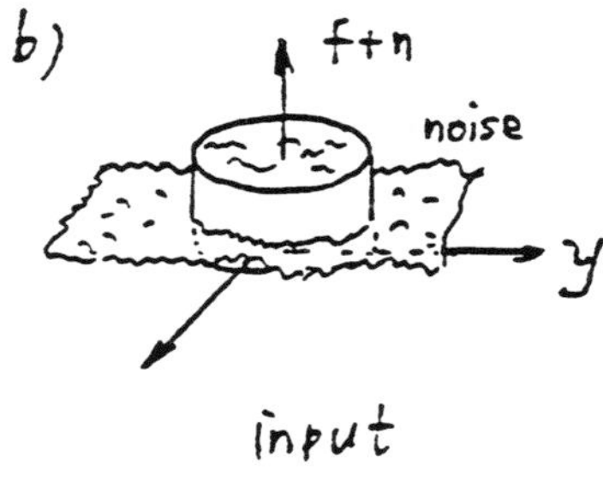

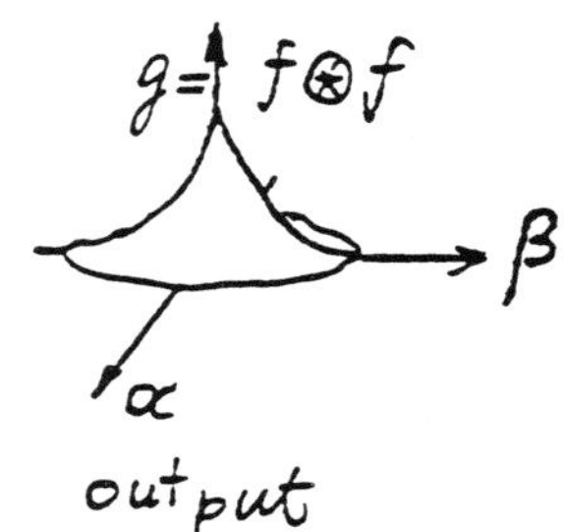

11.9

(a) The new input is $f(x-x_0, y-y_0) + n(x,y)$, its Fourier transform is $F(p,q)\, e^{-j(px_0+qy_0)} + N(p,q)$, see § 1.5.1.

$$|F(p,q) + N(p,q)| = |F(p,q)\, e^{-j(px_0+qy_0)} + N(p,q)|$$

So we call it shift invariant

(b) Immediately behind P_2, the light field is

$$[F(p,q)\, e^{-j(px_0+qy_0)} + N(p,q)]\, F^*(p,q)$$

$$= F(p,q)\, F^*(p,q)\, e^{-j(px_0+qy_0)} + N(p,q)\, F^*(p,q)$$

Thus the output is its inverse Fourier transform $f \circledast f_{(\alpha-x_0, \beta-y_0)} + 0 = R_{11}(\alpha-x_0, \beta-y_0)$

11.10

(a)

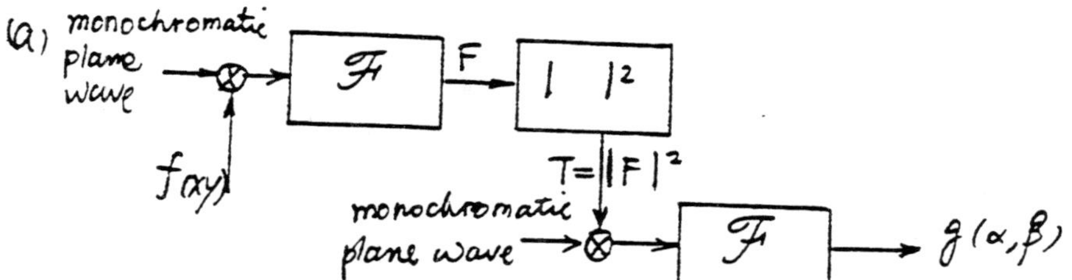

(b) $f(x,y) \rightarrow F(p,q)$, received by the square law converter,

the input to second lens is

$T = |F(p,q)|^2$, the output is the Fourier transform of T, $g(\alpha,\beta) = f(x,y) * f(x,y)$

(i.e. auto-correlation)

11.11

(a) $f(x,y) \to$ [$\mathcal{F}$] $\xrightarrow{F}$ (×) $\xrightarrow{G}$ [$\mathcal{F}^{-1}$] $\to g(\alpha,\beta)$

$H = i(p+q)$

$$G = HF = i(p+q)\,F(p,q)$$

The output is $g(\alpha,\beta) = \dfrac{\partial^2 f(\alpha,\beta)}{\partial\alpha\,\partial\beta}$

(b)

$$f = \text{rect}\left(\frac{\alpha}{a}, \frac{\beta}{b}\right), \quad \frac{\partial f}{\partial\alpha} = [\delta(\alpha-a)+\delta(\alpha+a)]\,\text{rect}\left(\frac{\beta}{b}\right)$$

$$\frac{\partial^2 f}{\partial\alpha\,\partial\beta} = [\delta(\alpha-a)+\delta(\alpha+a)][\delta(\beta-b)+\delta(\beta+b)]$$
$$= \delta(\alpha+a,\beta+b)+\delta(\alpha+a,\beta-b)+\delta(\alpha-a,\beta+b)+\delta(\alpha-a,\beta-b)$$

11.12

(a) The transmittance of the grating may be expressed as

$$t = \text{rect}\,\frac{x}{b} * \left(\text{rect}\,\frac{x}{L} \cdot \frac{1}{d}\,\text{comb}\,\frac{x}{d}\right)$$

where b is the width of groove; L is the side length; d is the grating constant, i.e., distance of neighboring grooves.

The Fourier spectrum of t is

$$T(f_x) = bL \operatorname{sinc} bf_x \cdot (\operatorname{sinc} Lf_x * \operatorname{comb} df_x),$$

where $f_x = \frac{p}{2\pi} = \frac{\xi}{\lambda f}$, we call comb df_x as multi-slits interference factor, $\frac{1}{d}$ is the interval of different diffraction orders. sinc bf_x is single-slit factor. it gives the envelope, sinc Lf_x decides the shape of each order, hence the resolution limit.

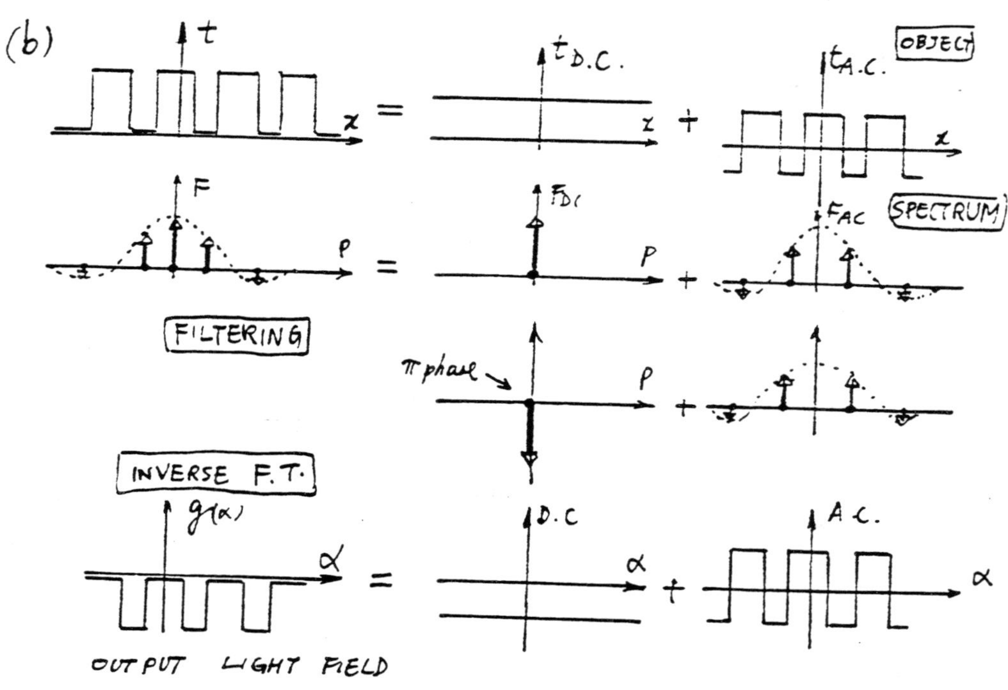

So the light intensity distribution is contrast reversed.

$I = |g|^2$

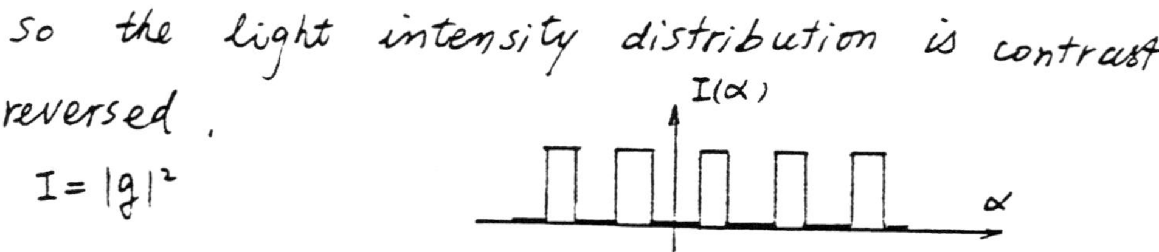

11.13

(a) There are two beams incident to the recording medium:

$u(\alpha,\beta) = \mathcal{F}[f(x,y)]$

$v(\alpha,\beta) = e^{jk\alpha\sin\theta}$

So the intensity distribution is $I(\alpha,\beta) = |u+v|^2$

$= |u|^2 + |v|^2 + u\, e^{-jk\alpha\sin\theta} + u^* e^{jk\alpha\sin\theta}$.

Neglecting the distribution of $\phi(u)$,

$I(\alpha,\beta) = K + |u|\, e^{-jk\alpha\sin\theta + j\phi(u)} + |u|\, e^{-j\phi(u) + jk\alpha\sin\theta}$

$= K_1 + K_2 \cos[k\alpha\sin\theta - \phi(u)] = K_1 + K_2\cos(k\alpha\sin\theta)$

$= K_1 + K_2 \cos\left(2\pi\alpha \frac{\sin\theta}{\lambda}\right)$

So the carrier frequency is $\frac{\sin\theta}{\lambda} = \frac{\sin 30^\circ}{6500\text{Å}}$

$= 0.5 \times \frac{1}{6500\times10^{-7}\,mm} = 769$ lines/mm

(b) The resolution of the recording medium should be at least 769 lines/mm, or even higher.

11.14

The transmittance of the ideal inverse filter $H(p) = \frac{1}{F(p)}$, for those locations where $F(p)=0$, $H(p)$ must be infinitive, so physically unrealizable.

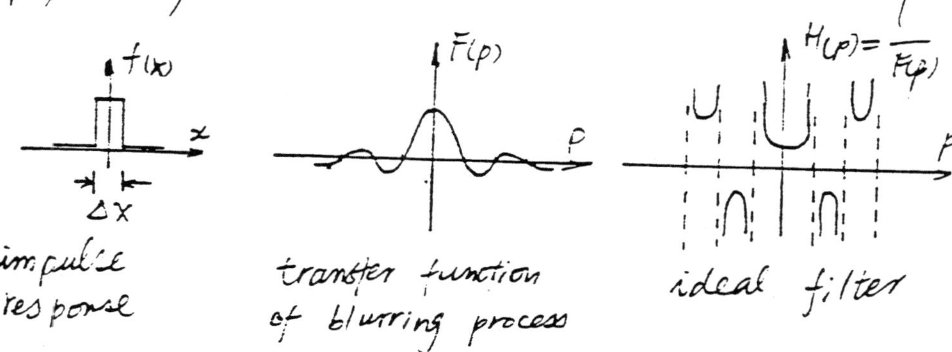

impulse response

transfer function of blurring process

ideal filter

The maximum transmittance of a practical filter is 1, never more than 1.

11.15

(a) No object is inserted in the input plane, that means uniform illumination. Its Fourier transform is a δ-function.

$H(p,q) = K_1 + K_2(e^{j\alpha_0 p + \phi} + e^{-j\alpha_0 p - \phi})$

Immediately behind the filter, there are

$K_1 \delta(p,q) + K_2 \delta(p,q) e^{j\alpha_0 p} + K_2 \delta(p,q) e^{-j\alpha_0 p}$

(Notice that the size of focused spot is actually larger than the spatial period of the implicit grating structure.)

So the output is $K_1 + K_2 + K_2$. If we neglect the lens size, a uniform distribution is observed. If the lens size is taken into account, then we see some overlapping.

(b)

11.16

(a) Input object is $f_1(x-\alpha_0, y)$ $f_2(x+\alpha_0, y)$

Its Fourier transform is

$F = F_1(p,q)\, e^{-jp\alpha_0} + F_2(p,q)\, e^{jp\alpha_0}$, the light intensity distribution on to the square law detecor is $|F|^2 = |F_1|^2 + |F_2| + 2|F_1F_2| \cos(2p\alpha_0 - \phi_{12})$

$= K_1 + K_2 \cos(2\pi\xi \frac{2\alpha_0}{\lambda f} - \phi_{12})$

Spatial frequency is $\frac{2\alpha_0}{\lambda f} = \frac{2\text{ cm}}{6000\text{Å}\ 50\text{ cm}}$

$= 66.7$ lines/mm

(b) The requirement is 66.7 lines/mm

11.17

(a) At the input plane, the light field is

$f(x,y) = f_1(x-x_0, y-y_0) + f_2(x,y)$, $x_0 = y_0 = 10$ mm,

The Fourier spectrum of $f(x,y)$ is

$F(p,q) = \mathcal{F}[f(x,y)] = F_1(p,q)\, e^{-j(px_0+qy_0)} + F_2(p,q)$

The corresponding intensity distribution is

$I(p,q) = |F(p,q)|^2 = F F^*$

$= |F_1|^2 + |F_2|^2 + F_1 F_2^*\, e^{-j(px_0+qy_0)} + F_1^* F_2\, e^{+j(px_0+qy_0)}$

where $p = \frac{2\pi\xi}{\lambda f}$, $q = \frac{2\pi\eta}{\lambda f}$

or,

$I(p,q) = |F_1|^2 + |F_2|^2 + |F_1||F_2| \cos[(px_0+qy_0) + \phi_2 - \phi_1]$

where $\phi_2(p,q)$, $\phi_1(p,q)$ are the phase distributions of $F_1(p,q)$ and $F_2(p,q)$, respectively.

(b) the reflectance of the square law converter

$R(p,q) = k\, I(p,q)$

the virtual light field on (p,q) plane is

$G = A R = A k\, I(p,q) = C\, I(p,q)$, where A is the amplitude of the readout beam. C arbitrary constant. On (α,β) plane, we obtain the inverse Fourier transform of $G(p,q)$

$g(\alpha,\beta) = \frac{1}{4\pi^2}\iint G(p,q)\, e^{j2\pi(p_1\alpha + q_1\beta)}\, dp_1 dq_1$,

$p_1 = \frac{2\pi\xi}{\lambda_1 f}$, $q_1 = \frac{2\pi\eta}{\lambda_1 f}$, λ_1 is the wavelength of the readout light. When $\lambda_1 = \lambda = 6000\,\text{Å}$

$$g(\alpha,\beta) = \frac{C}{4\pi^2}\iint [|F_1|^2 + |F_2|^2 + F_1F_2^* e^{-j(px_0+qy_0)} + F_1^*F_2 e^{j(px_0+y_0q)}] \cdot e^{j2\pi(p\alpha+q\beta)}\,dp\,dq$$

$$= C[f_1 \circledast f_1 + f_2 \circledast f_2 + f_1 \circledast f_2(\alpha - x_0, \beta - y_0) + f_2 \circledast f_1(\alpha + x_0, \beta + y_0)]$$

When $\lambda_1 = 4000\,\text{Å}$

$$g(\alpha,\beta) = \frac{C}{4\pi^2}\iint [|F_1(p,q)|^2 + |F_2(p,q)|^2 + F_1(p,q)F_2^*(p,q)\,e^{-j(px_0+qy_0)} + F_1^*(p,q)F_2(p,q)\,e^{j(px_0+qy_0)}]\, e^{j2\pi(p_1\alpha+q_1\beta)}\,dp_1\,dq_1$$

$$= \frac{C}{4\pi^2}\iint [|F_1(\tfrac{p_1}{m},\tfrac{q_1}{m})|^2 + |F_2(\tfrac{p_1}{m},\tfrac{q_1}{m})|^2 + F_1(\tfrac{p_1}{m},\tfrac{q_1}{m})F_2^*(\tfrac{p_1}{m},\tfrac{q_1}{m})\,e^{-j(p_1\frac{x_0}{m}+q_1\frac{y_0}{m})} + F_1^*(\tfrac{p_1}{m},\tfrac{q_1}{m})F_2(\tfrac{p_1}{m},\tfrac{q_1}{m})\,e^{j(\frac{p_1x_0}{m}+\frac{q_1y_0}{m})}]\, e^{j2\pi(p_1\alpha+q_1\beta)}\,dp_1\,dq_1$$

$$= C\{f_1(m\alpha,m\beta) \circledast f_1(m\alpha,m\beta) + f_2(m\alpha,m\beta) \circledast f_2(m\alpha,m\beta)$$
$$f_1(m\alpha,m\beta) \circledast f_2[m(\alpha - x_0),\, m(\beta - y_0)] + f_2(m\alpha,m\beta) \circledast f_1[m(\alpha + \tfrac{x_0}{m}),\, m(\beta + \tfrac{y_0}{m})]\}$$

where $m = \frac{p_1}{p} = \frac{\lambda}{\lambda_1}$

So the scale changed to a smaller one.

11.18

a) There is no requirement on the extended light source, this is the major advantage of source-encoding technique.

As for the encoding mask, it is actually a grating, or multi-slits. It can be expressed as

$$t_M = \text{rect}\frac{x_0}{b} * \sum_n \delta(x_0 - nd)$$

For simplicity, assume the focal length of the collimating lens $f_0 = f_1$, the light distribution on P_1 is the convolution of t_M and the Fourier spectrum of input object plus grating G_1, so b should be less than the main lobe of the input object.

DC component can be decided as follows, the background of the input object is

$$f(x) = \text{rect}\frac{x_1}{l},$$

its Fourier spectrum $F(\alpha) = l \,\text{sinc}\, l\frac{\alpha}{\lambda f}$

$a = \frac{2\lambda f}{l}$, for red light

$$a = \frac{2 \times 700\ nm \times 500\ mm}{5\ mm} = 0.14\ mm$$

$b \ll a$, say $0.07\ mm$.

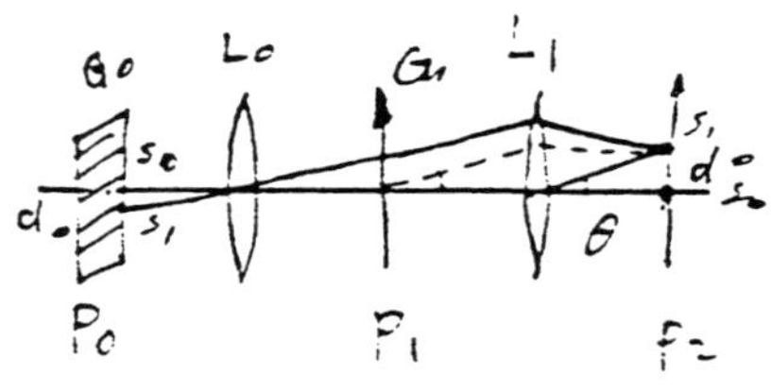

the spacing between slit #0 and slit #1, is d.

the first diffraction order of G_1 should be at the location where the image of S_1 is.

$$d_1 \sin\theta = \lambda$$

$$d_1 \frac{d_0}{f} = \lambda, \quad \therefore \quad d_0 = \frac{\lambda f}{d_1}$$

So d_0 depends upon the wavelength and the modulation grating d_1.

b) $f_x = 2$ lines/mm

$\alpha = \lambda f f_x$

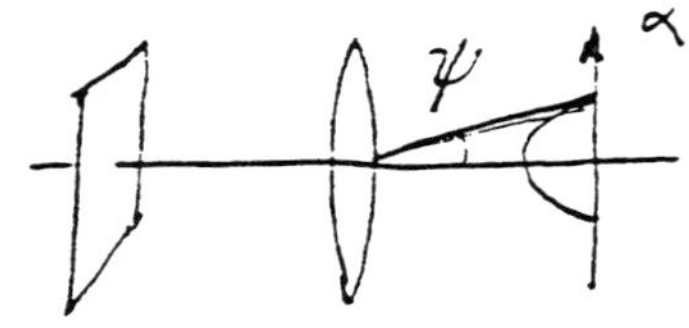

The size of H_N is 2α

$= 2\lambda f f_x$

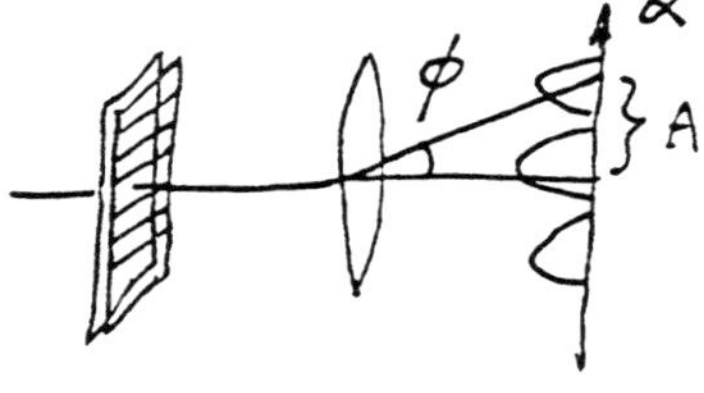

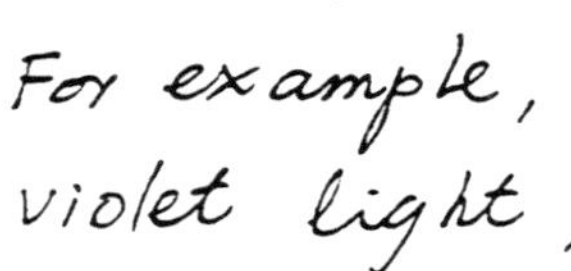

For example,

violet light,

$$2\alpha_v = 2 \times 350\,nm \times 500\,mm \times 2\,/mm$$

$$= 0.7\ mm$$

red light $2\alpha_r = 2 \times 700\,nm \times 500\,mm \times 2/mm$

$= 1.4\ mm.$

to avoid the overlapping of neighboring orders, $A \geq 2\lambda f f_x$

$$A = \lambda f f_{Mod}$$

$$\therefore f_{Mod} \geq 2 f_x \geq 4 \text{ lines/mm}$$

where f_{Mod} is the frequency of modulating grating. This is for quasi-monochromatic processing.

For multicolor processing, the over-lapping will be severer at higher orders, $f_{Mod} = 4N$ lines/mm, N is the number of filters.

11.19

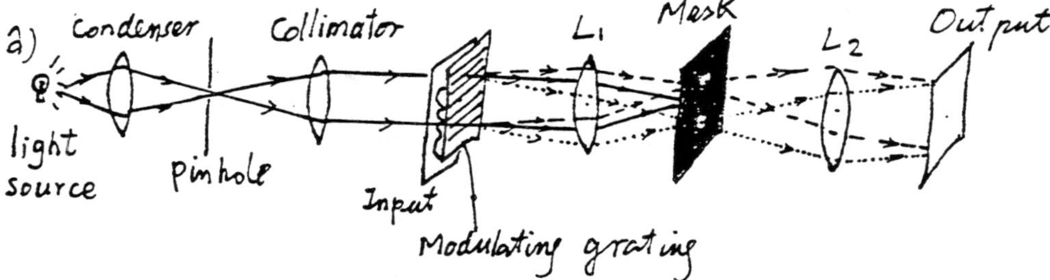

b)

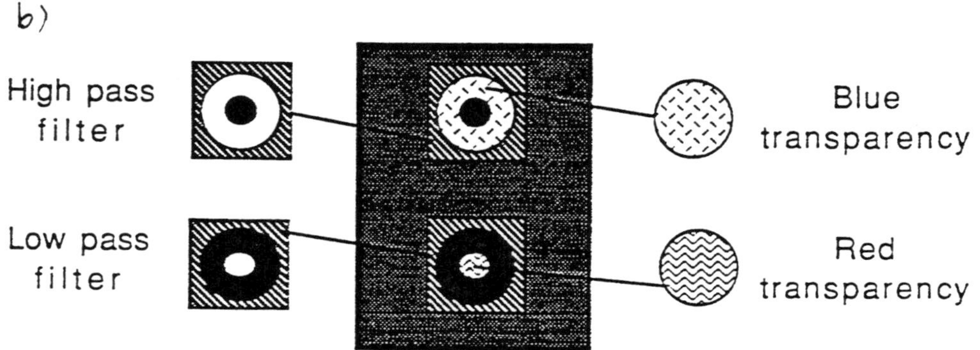

the Compoud Filter for
Frequency Pseudocolor Encoding

11.20

(a) The spectrum components of the white light source may be expressed as $I_i(\lambda_i)$, $i=1,2,\cdots$. Each component yields optical Fourier transform. The monochromatic illuminating amplitude is $A_i(\lambda_i)$, $A_i=\sqrt{I_i}$,

The input is always $t(x,y)$, the light field behind the input plane is

$$f_i(x,y,\lambda_i)=A_i(\lambda_i)\,t(x,y)$$

$$=A_i(\lambda_i)[K+t_r(x,y)\cos(p_0x)+t_g(x,y)\cos(p_0y)+t_b(x,y)\cos(2p_0y)].$$

Its Fourier transform is

$$F_i(p_i,q_i)=\mathcal{F}(x,y,\lambda_i)$$

$$= 2A_i(\lambda_i)\left[\frac{\delta(p_i, q_i)}{2} + T_r(p_i - p_0, q_i) + T_r(p_i + p_0, q_i) + T_g(p_i, q_i - p_0) + T_g(p_i, q_i + p_0) + T_b(p_i, q_i - 2p_0) + T_b(p_i, q_i + 2p_0)\right]$$

Please notice, different wavelengths yields different p_i, q_i. ($p_i = \frac{\alpha}{\lambda_i f}$, $q_i = \frac{\beta}{\lambda_i f}$), hence different scale.

There is no coherence between different spectrum components. So the compound light intensity distribution on (α, β) plane is the superposition of Fourier spectra of different wavelength at different scales.

$I(\alpha, \beta) = \sum |F_i|^2$.

(b) the light distribution on (α, β) plane and the color transparent filters

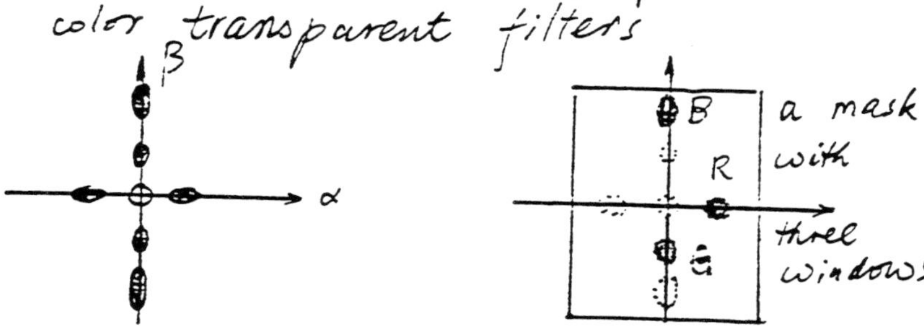

(c) Only $T_r(p_r - p_0, q_r)$, $T_g(p_g, q_g + p_0)$ and $T_b(p_b, q_b - 2p_0)$ can pass, so the resultant output is $\frac{1}{4}[I_r(t_r)^2 + I_g(t_g)^2 + I_b(t_b)^2]$

11.21 / The sequence of process :

1/ Take the input (CCD)

2/ Display input & reference (SLM1)

3/ Take the JTPS (CCD1)

4/ Display the JTPS (SLM2)

5/ Take the correlation (CCD2)

Therefore, not including the data transfer time from and to the computer, the

$$\text{process time} = t_{CCD} + t_{SLM_1} + t_{CCD_1} + t_{SLM_2} + t_{CCD2}$$

$$= 5 \times \frac{1}{60} \text{ sec} = \frac{1}{12} \text{ sec}$$

∴ The duty cycle is 12 correlations/sec.

11.22 a) Process sequence :

1/ Take input (CCD)

2/ Display input & reference (SLM/LCTV)

3/ Take JTPS with CCD1

4/ Display JTPS

5/ Take the correlation (CCD1)

process time = $5 \times \frac{1}{60}$ sec = $\frac{1}{12}$ sec

Duty cycle = 12 correlations/sec.

b) It looks like this system only has advantages since it has the same duty cycle with fewer number of components (only 1 SLM compared to 2 and only 2 CCDs instead of 3).

However if we look more closely we can see that we can synchronize the CCD1 and SLM2 of prob 10.21 so that it only needs $4 \times \frac{1}{60}$ sec = $\frac{1}{15}$ sec fof each process.

Other thatz that the transfer time from and to computer memory normally cannot be ignored, even sometimes needs much more time than the other process. when this is true, the previous set up in prob 10.21 is much faster.

In conclusion we see the need of compromise of faster system and less complicated set up.

11.24 Using eqn. (11.53)

(a) $\left|\frac{\Delta\lambda}{\lambda_1}\right| < \frac{\lambda_1}{\eta d} = \frac{488 \times 10^{-9}}{2.55 \times 10^{-3}}$

$\simeq 1.91 \times 10^{-4}$

OR

$\Delta\lambda_{max} \simeq 0.093$ nm

(b) The wavelength spread (bandwith) of the laser diode needs to be much smaller than $\Delta\lambda_{max}$, practically about one-fifth of $\Delta\lambda_{max}$

11.25 $f = 1$ m using (11.55)

(a) $|\Delta x(x_1 - \Delta x)|_{max} < \lambda_1 f^2 / nd$

$|\Delta x(x_1 - \Delta x)|_{max} < 488 \times 10^{-6} \frac{(1000)^2}{2.25 \times 1} (mm)^2$

$|\Delta x(x_1 - \Delta x)|_{max} < 217 (mm)^2$

(b) Object is limited to 1.5 mm in size.
We assume the reference is centered $\rightarrow x_1\ max = \frac{1.5\ mm}{2}$

$|\Delta x(x_1 - \Delta x)|_{max} = |-\Delta x(0.75 + \Delta x)|$

$= 0.75|\Delta x| + (\Delta x)^2 < 217 (mm)^2$

$|\Delta x|_{max} \simeq 14.36$ mm

11.23

a) Process time $= t_{CCD} + t_{SLM_1} + t_{SLM_2} + t_{CCD2}$

$$= 4 \times \frac{1}{60} \text{ sec} = \frac{1}{15} \text{ sec}$$

duty cycle = 15 correlations/sec.

If we can send the input & the reference filter to SLM1 and SLM2 simultaneously we can reduce the process time to:

$$t_{CCD} + t_{SLM} + t_{CCD_2} = \frac{1}{20} \text{ sec}$$

→ duty cycle = 20 correlations/sec.

b) This result is faster compared to the two preceeding set ups (10.21 and 10.22) but it has to be noted that the filters need to be calculated beforehand. In the previous problems there is no need to calculate the filter because they are the original reference (not in frequency domain)

c) As has been stated above, the JTCs does not need filter synthesis but the Vander Lugt correlator does. Normally, the filter in spatial frequency domain (for Vander Lugt) needs higher resolution than its original form in spatial domain.

Chapter 12

12.1 (a) $\phi_c = \sin^{-1}\left(\frac{n_a}{n_g}\right) = \sin^{-1}\left(\frac{1}{1.48}\right) = 42.5°$

(b) $\theta_1 = 90° - \phi_c = 47.5°$

$n_1 \cdot \sin\theta_1 = 1.48 \cdot \sin 47.5° = 1.09 > 1.00$

$\therefore$ acceptance angle is $90°$.

(c) $NA = n_0 \sin\theta_0 = 1$

12.2 (a). $\phi_c = \sin^{-1}\left(\frac{n_2}{n_1}\right) = \sin^{-1}\left(\frac{1.43}{1.46}\right) = 78.36°$

(b). $\theta_1 = 90° - \phi_c = 11.64°$

$\theta_a = \sin^{-1}\left(\frac{n_1}{n_w} \cdot \sin\theta_1\right) = 11.88°$

(c) $NA = n_w \cdot \sin\theta_a = 0.29$

12.3 $\theta_a = \sin^{-1}\left(\frac{\sqrt{n_1^2 - n_2^2}}{n_0}\right) = \sin^{-1}\left(\sqrt{1.46^2 - 1.45^2}\right) = 9.82°$

$\sin\theta_1 = \frac{\sin\theta_a}{n_1}$, $\theta_1 = 6.71°$

For single mode transmission:

$$\sin\theta_1 < \frac{(1+1)\lambda}{2d}$$

$$\therefore d < \frac{\lambda}{\sin\theta_1} = 6 \ \mu m$$

12.4 $NA = \sin\theta_a = 0.17$

$D = NA \cdot f = 0.17 f$

12.5 $\phi_c = \sin^{-1}\left(\frac{\eta_2}{\eta_1}\right) = \sin^{-1}(0.99) = 81.9°$

$\theta_1 = 90° - \phi_c = 8.9°$

$NA = \eta_1 \cdot \sin\theta_1 = 0.21$

12.6 $\Delta T = \eta_1 \left(1 - \frac{\eta_2}{\eta_1}\right) \frac{L}{c}$

$= \eta_1 (1 - 0.98) \frac{3 \times 10^3}{3 \times 10^8}$

$= 6 \times 10^{-7} \eta_1 \quad (sec)$

If $\eta_1 = 1.5$,

$\Delta T = 9 \times 10^{-7}$ sec $= 0.9$ μsec.

12.7 $\phi_c = \sin^{-1}\left(\frac{\eta_2}{\eta_1}\right) = \sin^{-1}\left(\frac{1.46}{1.48}\right) = 1.406$ rad.

$d < \frac{\lambda}{\frac{\pi}{2} - \phi_c} = 5.16$ μm.

12.8 $\phi_c = \sin^{-1}\left(\frac{\eta_2}{\eta_1}\right) = 1.338$ rad.

$d < \frac{\lambda}{\frac{\pi}{2} - \phi_c} = 3.65$ μm

12.9 (a) $\Delta = \frac{\eta_1^2 - \eta_2^2}{2\eta_1^2} = \frac{(\eta_1 + \eta_2)(\eta_1 - \eta_2)}{2\eta_1^2} \approx \frac{2\eta_1(\eta_1 - \eta_2)}{2\eta_1^2}$

$= \frac{\eta_1 - \eta_2}{\eta_1}$

(b) $NA = \sqrt{\eta_1^2 - \eta_2^2} \approx \sqrt{2\eta_1^2 \Delta} = \eta_1 \sqrt{2\Delta}$

12.10 $\eta_1 = 1.45$

$\eta_2 = 0.99\,\eta_1 = 1.4355$

$J = \frac{2\pi}{\lambda} \cdot \frac{d}{2} \sqrt{\eta_1^2 - \eta_2^2} = 75.6$

$(n+1) = \frac{2d\sin\theta_1}{\lambda}, \quad \theta_1 = \sin^{-1}[90° - \phi_c] = 8.9°$

$\therefore N = \frac{2 \times 100 \cdot \sin 8.9}{0.85} - 1 = 34$

12.11 $d = 80\ \mu m, \quad NA = 0.2, \quad \lambda = 0.7\ \mu m, \quad \alpha = 2$

$J = \frac{2\pi}{\lambda} \cdot \frac{d}{2} \sqrt{\eta_1^2 - \eta_2^2} = 71.8$

$M_g = \frac{\alpha}{\alpha + 2} \left(\eta_1 k \cdot \frac{d}{2}\right)^2 \Delta = \frac{\alpha}{\alpha + 2} \cdot \frac{J^2}{2} = 1289$

12.12 $\phi_c = \sin^{-1}\left(\frac{\eta_2}{\eta_1}\right) = 79.74°, \quad \theta_1 = 90° - \phi_c = 10.26°$

$d = \frac{(n+1)\lambda}{2\sin\theta_1} = \frac{1001 \times 0.6\ \mu m}{2 \cdot \sin 10.26°} = 1.68\ mm$

12.13 (a) $\Delta = \frac{\eta_1 - \eta_2}{\eta_1} = 3.3\%$

(b) $\theta_1 = 90° - \sin^{-1}\left(\frac{\eta_2}{\eta_1}\right) = 14.84°$

$\lambda_{min} = 2d \cdot \sin\theta = 2.56\ \mu m$

12.14 $d = 10\ \mu m, \quad \lambda = 2.56\ \mu m$

$\sin\theta_1 = \frac{\lambda}{2d} = 0.128, \quad \theta_1 = 7.35°$

$\phi_c = 90° - \theta_1 = 82.65°$

$\eta_2 = \eta_1 \cdot \sin\phi_c = 1.488$

12.15 $\sin\theta_1 = \frac{\sin\theta_a}{\eta_1} = \frac{\sin 7.5°}{1.52}$

$\theta_1 = 4.93°$

$\phi_c = 90° - \theta_1 = 85.07°$

$\eta_2 = \eta_1 \cdot \sin\phi_c = 1.514$

12.16 $J = \frac{2\pi}{\lambda} \cdot \frac{d}{2} \cdot \sqrt{\eta_1^2 - \eta_2^2} = 32.95$

$M_g = \frac{\alpha}{\alpha + 2} \cdot \frac{J^2}{2} = 257$

12.17 $\frac{10000}{781} = 12.8$

∴ At least 13 wavelength channels are required.

12.18 Total loss : $-0.5 \times 10 = -5\,dB$.

$$I_{out} = 10^{-0.5} \cdot I_i = 31.6\%\ I_i$$

12.19 $10 \log (50\%) = -3\ dB$

$$\frac{-3\ dB}{-0.3\ dB/km} = 10\ Km.$$

∴ distance between repeaters is 10 Km.

12.20 No solution is provided.

12.21 $\Delta p = \frac{2\pi \cdot \Delta l \cdot n}{\lambda} = \frac{2\pi \times 0.3 \times 1.5}{0.67} = 1.34\ \pi$.

∴ detected intensity changes from minimum to maximum and then reduced a little.

12.22 No solution is provided. Discussion is encouraged

12.23 No solution is provided. Discussion among students is encouraged.

12.24

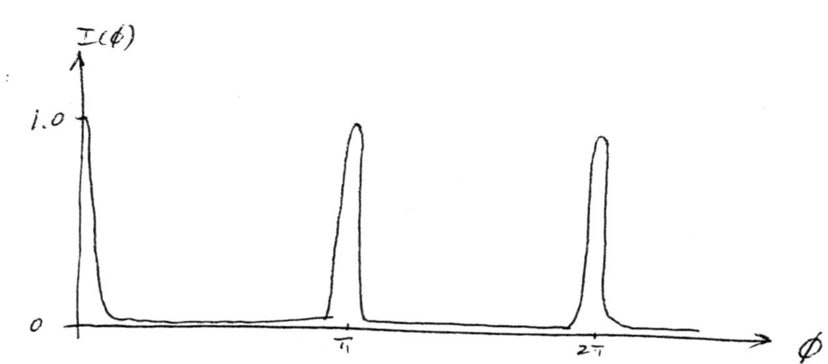

12.25 No solution is available. Discussion among students is encouraged.

12.26 If $\omega_s = \omega_L$,

$$I(t) = I_S + I_L + 2\sqrt{I_S I_L}\cos[\phi_s - \phi_L]$$

It changes with phase of the signal ϕ_s.

This is called homodyne detection

If ϕ_s does not change, $I(t)$ is a constant.

12.27

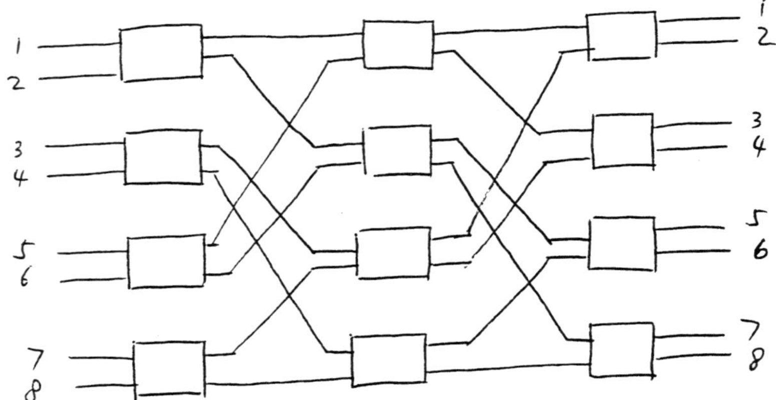

12.28

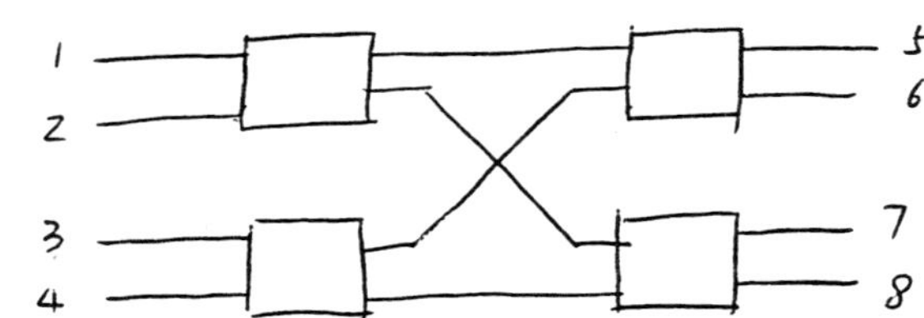